Felix Ntui Ayuk

Surface plasmons resonance spectroscopy

Felix Ntui Ayuk

Surface plasmons resonance spectroscopy

Surface Plasmons resonance spectroscopy and its application to sensor devices. A novel approach with x-ray spectroscopy

Südwestdeutscher Verlag für Hochschulschriften

Impressum/Imprint (nur für Deutschland/only for Germany)
Bibliografische Information der Deutschen Nationalbibliothek: Die Deutsche Nationalbibliothek verzeichnet diese Publikation in der Deutschen Nationalbibliografie; detaillierte bibliografische Daten sind im Internet über http://dnb.d-nb.de abrufbar.
Alle in diesem Buch genannten Marken und Produktnamen unterliegen warenzeichen-, marken- oder patentrechtlichem Schutz bzw. sind Warenzeichen oder eingetragene Warenzeichen der jeweiligen Inhaber. Die Wiedergabe von Marken, Produktnamen, Gebrauchsnamen, Handelsnamen, Warenbezeichnungen u.s.w. in diesem Werk berechtigt auch ohne besondere Kennzeichnung nicht zu der Annahme, dass solche Namen im Sinne der Warenzeichen- und Markenschutzgesetzgebung als frei zu betrachten wären und daher von jedermann benutzt werden dürften.

Coverbild: www.ingimage.com

Verlag: Südwestdeutscher Verlag für Hochschulschriften GmbH & Co. KG
Dudweiler Landstr. 99, 66123 Saarbrücken, Deutschland
Telefon +49 681 37 20 271-1, Telefax +49 681 37 20 271-0
Email: info@svh-verlag.de

Approved by: Siegen, Universität, Diss., 2010

Herstellung in Deutschland:
Schaltungsdienst Lange o.H.G., Berlin
Books on Demand GmbH, Norderstedt
Reha GmbH, Saarbrücken
Amazon Distribution GmbH, Leipzig
ISBN: 978-3-8381-2837-5

Imprint (only for USA, GB)
Bibliographic information published by the Deutsche Nationalbibliothek: The Deutsche Nationalbibliothek lists this publication in the Deutsche Nationalbibliografie; detailed bibliographic data are available in the Internet at http://dnb.d-nb.de.
Any brand names and product names mentioned in this book are subject to trademark, brand or patent protection and are trademarks or registered trademarks of their respective holders. The use of brand names, product names, common names, trade names, product descriptions etc. even without a particular marking in this works is in no way to be construed to mean that such names may be regarded as unrestricted in respect of trademark and brand protection legislation and could thus be used by anyone.

Cover image: www.ingimage.com

Publisher: Südwestdeutscher Verlag für Hochschulschriften GmbH & Co. KG
Dudweiler Landstr. 99, 66123 Saarbrücken, Germany
Phone +49 681 37 20 271-1, Fax +49 681 37 20 271-0
Email: info@svh-verlag.de

Printed in the U.S.A.
Printed in the U.K. by (see last page)
ISBN: 978-3-8381-2837-5

Copyright © 2011 by the author and Südwestdeutscher Verlag für Hochschulschriften GmbH & Co. KG and licensors
All rights reserved. Saarbrücken 2011

Surface Plasmons Resonance Spectroscopy and its application to sensor devices: A novel approach with the combination of X-ray spectroscopy

Dissertation

zur Erlangung des Grades eines Doktors der Naturwissenschaften

(Dr.rer.nat) vorgelegt von:

Dipl.-Ing.(FH)
Felix Ntui Ayuk

eingereicht beim Fachbereich -7, Physik der Universität Siegen

Siegen 2010

Die vorliegende Arbeit wurde in den Zeitraum von Oktober 2006 bis Januar 2011 im Arbeitskreis von Prof. Ulrich Pietsch der Universität Siegen durchgeführt.

1. Gutachter: Prof. Dr. Ulrich Pietsch
2. Gutachter: Prof. Dr. Holger Schönherr

Tag der Disputation: 24.01.2011

Abstract

English..5
Deutsch..7

1 General Introduction

1.1 State of the Art in Biosensors..9
1.2 Surface Plasmons based sensing technique..........................11
1.3 Aim of the Thesis..13

2 Theory

2.1 Evanescence wave theory..14

2.1.1 Plane wave at interfaces...15
2.1.1.1 Maxwells equation and plane wave....................................15
2.1.1.2 Fresnels equations...16
2.1.1.3 Total internal reflection..18

2.1.2 Surface Plasmons polaritons..19
2.1.2.1 Solving Maxwells equation...20
2.1.2.2 Excitation of Surface Plasmons..24
2.1.2.2.1 Prism coupling...25
2.1.2.2.2 Bare metal reference...27
2.1.2.2.3 Thin additional film..30

3 X-Ray Reflectivity

3.1 Basic principles of X-ray reflectivity..32
3.2 The critical angle of reflection...33
3.3 Reflected intensity from ideally flat surface...........................33

4 Experimental Methods

4.1 Sample Characterisation Techniques.....................................41

4.1.1 Surface Plasmons Spectroscopy..41

4.1.1.1 Measurement Modes..41

4.1.1.1.1 Angular Dependent Measurments..................................41
4.1.1.1.2 Time Dependent Measurements....................................43

4.1.1.2 Measurement Setup..46

4.1.1.2.1 Surface Plasmons Specstroscopy...46
4.1.1.2.2 X-Ray Reflectometry..47

4.2 Sample Preperation Techniques...48

4.2.1 Thermal Everporation of Metal Layers..48
4.2.2 Self Assembled Monolayers on Metal..49
4.2.3 Drop-Casting of Polymer...50

5 Static Measurements

5.1 Detection of Alkanes, Alkynes Alcohol and organic acids.............................51
5.2 Effect of wavelength on the resonance angle...53
5.3 Simultaneous detection of thickness and dielectric constant of self assembled monolayer...55

6 Dynamic Measurement

6.1 Thermal Dynamic Diffusion of Gold Colloids through P3HT........................59

6.1.1 Theoretical consideration...62

6.1.1.1 Simulation of X-Ray Measurements using the simulation program XOP2.........68
6.1.1.2 Simulation of Surface Plasmons curves using X-ray model.........................79
6.1.1.3 Discusion..86

6.2 Non Diffusion of Gold Colloids on P3HT...90

6.2.1 Theoretical consideration...90

6.2.1.1 Simulation of X-Ray Measurements using XOP2....................................95
6.2.1.2 Simulation of Surface Plasmons curves using X-ray model........................100
6.2.1.3 Discusion 101

7 Conclusion..103

8 Biography...105

9 Acknowledgements..109

10 Affidavit 110

Abstract:

Almost 100 years ago, R. W. Wood observed strong, angular dependent variations in the intensity of light that was reflected from an optical metal grating. This effect was due to the interaction of light with a fundamental excitation of a metal-dielectric interface. This excitation is characterised by a charge density oscillation in the metal, which is accompanied by an electromagnetic field that extends in both media. Since the energy is confined to the vicinity of the metal surface, and the conduction electrons of a metal can be treated as plasma, this excitation is called the surface plasmons resonance.

For quite a long period of time, not much progress was made in this field until the availability of new theoretical and experimental techniques triggered growing interest in the optics of metallic thin film.

In this work, a self made surface plasmons resonance set-up was constructed using the Kretschmann's configuration. Using the self made device, diverse static measurements were carried out to verify the credibility of the sensor.

The bulk behavior of polymers has been investigated extensively since the discovery of their (commercial) value in many applications during the 1940s. Thus, nowadays the relation between molecular structure and macroscopic behaviour is fairly well understood.

However, since molecular interactions always show changes in thicknesses and refractive indices, by combining the device with x-ray spectroscopy would guarantee the independent monitoring of thickness and electron density variation during such interactions. In this way the self made device was then adapted to our home x-ray device and simultaneous measurements of SPR and x-ray reflectivity were also carried out on a polymer film.

Using both devices simultaneously we studied the diffusion of gold colloids through a thin layer of polymer initiated by an external perturbation, heat. By varying the temperature of the sample stepwise from 150°C to 200°C, a vertical distribution of gold colloids within the polymer was created. This vertically induced electron density variation was recorded independently using the x-ray reflectivity device after each annealing step. Simultaneously the optical changes brought about by the vertically induced electron density were also recorded using the SPR sensor device. By simulating the results obtained from the x-ray reflectivity and the subsequent calculation of the electron density within the polymer layer, we could later verify the responds recorded by the SPR response.

Finally colloids with a diameter greater than the thickness of the polymer film were also brought onto the surface of the polymer layer and the sample also annealed from 150°C to 220°C. Although this time there was no diffusion of the colloids into the polymer, the submerged part of the colloids into the polymer also created an electron density variation within the polymer layer. This was later modelled and simulated. This generated electron density variation as a result of this submersion of the colloids into the polymer was later seen to be responsible for the characteristic movement of the SPR response towards higher angles. The simultaneous study of a dynamic process using both spectroscopic processes has never been made up till date.

Oberflächenplasmonresonanzspektroskopie und ihre Anwendungen in Sensorgeräten: Ein neuartiges Model mit der Kombination der Röntgenspektroskopie.

Kurzfassung:

Vor fast 100 Jahren, R. W. Wood beobachtete starke, winkelabhängige Schwankungen in der Intensität des Lichtes, das von einer optischen Metallvergitterung reflektiert wurde. Diese Effekten sind wegen der Wechselwirkung des Lichtes mit einer grundsätzlichen Erregung einer metall-dielektrischen Schnittstelle. Diese Erregung wird durch eine Ladundsdichte-Schwingung im Metall charakterisiert, das durch ein elektromagnetisches Feld begleitet wird, das sich in beiden Medien ausstreckt. Da die Energie zur Umgebung der Metalloberfläche beschränkt wird, und die Leitungselektronen eines Metalls als ein Plasma behandelt werden können, wird diese Erregung die Oberflächeplasmonresonanz genannt. Seit einem langen Zeitraum wurden nicht viele Fortschritte in diesem Feld gemacht bis die Verfügbarkeit von neuen theoretischen und experimentellen Techniken Interesse an der Optik des metallischen dünnen Schichten ausgelöst hat.

In dieser Dissertation wurde ein selbst gebauter Sensor für die Erregung von Oberflächenplasmonen nach der Konfiguration von Kretschmann gebaut. Mit dem selbst gebauten Gerät wurden verschiedene statische Massungen ausgeführt, um die Vertrauenswürdigkeit des Sensors nachzuprüfen.

Aufgrund der Tatsache dass Molekular Wechselwirkungen immer Änderungen in der Dicke, und Brechungsindizes nachweisen, durch der Verbindung dieses Gerät mit der Röntgenstrahl-Spektroskopie Gerät, würden Dicke und Elektrondichte-Schwankung während solcher Wechselwirkungen mitverfolgt. Mit beide Messverfahren gleichzeitig, studierten wir die Verbreitung von Goldkolloiden durch eine dünnen Schicht des Polymers P3HT. Die Temperatur der Probe wurde schrittweise von 150°C bis 200°C erhöht. Dadurch wurde eine vertikale Verteilung von Goldkolloiden innerhalb des Polymers geschaffen. Diese vertikale Elektrondichte-Schwankung wurde unabhängig mit dem Röntgenstrahl-Gerät nach jedem schritt der Temperaturerhöhung gemessen. Gleichzeitig wurden die optischen Änderungen, die durch die vertikale Elektrondichteänerungen veranlasst wurde, mit dem Oberflächenplasmonresonanz Sensorgerät gemessen. Durch die Simulation der Ergebnisse der Röntgenstrahl-Reflexionsvermögen und der nachfolgenden Berechnung der Elektrondichte innerhalb der Polymerschicht, konnten wir später die durch das Oberflächenplasmonresonanzgerät registrierten Antworten nachprüfen.

Schließlich wurden Kolloide mit einem Durchmesser größer als die Dicke der Polymerschicht auch auf die Oberfläche der Polymerschicht gebracht, und die Probe wurde von 150°C bis 220°C erwärmt. Obwohl es keine Verbreitung der Kolloide ins Polymer stattfand, schuf der untergetauchte Teil der Kolloide ins Polymer auch eine Elektrondichte-Schwankung innerhalb der Polymerschicht. Das wurde modelliert und simuliert. Wie man später herausfand, war diese erzeugte Elektrondichte-Schwankung infolge dieses Untertauchens der Kolloide ins Polymer für die charakteristische Verschiebung der Peak des Oberflächenplasmonresonanz zu grossere Winkeln. Die gleichzeitige Untersuchung eines dynamischen Prozesses bei Verwendung beide spektroskopischen Prozessen, ist bis dato nie gemacht worden.

1 General Introduction

1.1 State of the arts in Biosensors and future trends

A biosensor is defined as an analytical device which contains a biological recognition element immobilised on a solid surface and a transduction element which converts analytes binding events into a measurable signal [1, 5, 6]. Biosensors use the highly specific recognition properties of biological molecules to detect the presence of binding partners usually at very low concentrations. Biological recognition can surpass any man-made device in sensitivity and specificity. This specificity permits very similar analytes to be distinguished from one another by their interaction with immobilised bio-molecules. Biosensors are valuable tools for fast and reliable detection of analytes and have reached an importance for scientific, bio-medical and pharmaceutical applications [7, 8].

The advantages that are offered by the ideal biosensor over other forms of analytical techniques are:

1. high sensitivity and selectivity,
2. low detection limits,
3. good reproducibility,
4. rapid response, reusability of device, ease of fabrication and application
5. possibility of miniaturization, ruggedness and low fabrication cost.

By immobilising the bio recognition element on the sensor surface, one gains the advantage of reusability of the device due to the ease of separating bound and unbounded species. The mere presence of the analyte itself does not cause any measurable signal from the sensor but the selective binding of the analyte of interest to the biological component. The latter is coupled to a transducer which responses to the binding of the bio molecule[9, 10]. By simple washing steps, the non- specifically bound molecule may be removed. The evanescence wave technique even makes these washing steps redundant. These techniques are relatively insensitive to the presence of analytes in the bulk solution. The three most widely used surface sensitive transduction processes are: piezo(electric), electrochemical and optical detectors. While electrochemical sensors response to changes in ionic concentrations, redox potential or electron transfer rates or electron density upon analyte binding, piezoelectric sensors monitor changes in the absorbed mass on the sensor surface. A large number of optical biosensors are based on the principle of fluorescence, chemo-illuminescence or absorption spectroscopy. Surface sensitive techniques provide a vital link both for the understanding of bio molecular recognition and the development of bio sensors. Indeed surfaces and cell surfaces in particular are involved in many important biological functions via the cell surface its self (the recognition of foreign bodies by special receptors located on the cell surface for example) or across the cell membrane (as in the signal transduction from one neurone to another involving complex membrane receptor proteins). These interfaces are central to a variety of biochemical and biophysical processes: triggering of cellular responses by neurotransmitters binding, blood coagulation of foreign substances, cellular mobility etc. In addition, surface sensitive techniques bring an advantage over bulk techniques in that they provide real time binding data. By immobilising one of the partners of the binding process on the surface of the transducer, the binding of the complement can be followed unperturbed by the presence of free molecules on the bulk.This eliminates the need for lengthily and perturbing separation steps that are required in most bulk techniques. The techniques that provide surface sensitivity as well as being non- destructive and giving in situ responses can be classified by the method of detection on which they are based into the following:

1. Electrical: impedance spectroscopy
2. Acoustic: piezoelectric waveguides
3. Optical: ellipsometry and reflectometric interference, attenuated total internal reflection, infrared specstroscopy, surface plasmons resonance, total internal reflection and optical waveguides.

This research will be based partly on the principle of surface plasmons resonance making use of the principle of attenuated total internal reflection.

1.2 Surface plasmons based sensing technique:

Surface plasmon resonance (SPR) is a phenomenon occurring at metal surfaces (typically gold and silver) when an incident light beam strikes the surface at a particular angle [11, 31-34]. Depending on the thickness of a molecular layer at the metal surface, the SPR phenomenon results in a graded reduction in intensity of the reflected light. Biomedical applications take advantage of the exquisite sensitivity of SPR to the refractive index of the medium next to the metal surface, which makes it possible to measure accurately the adsorption of molecules on the metal surface and their eventual interactions with specific ligands. The last ten years have seen a tremendous development of SPR use in diverse applications.

The technique is applied not only to the measurement in real time of the kinetics of ligands receptor interactions and to the screening of lead compounds in the pharmaceutical industry, but also to the measurement of DNA hybridization, enzyme- substrate interactions, in polyclonal antibody characterization, protein conformation studies, diffusion of colloids through cell membrane and label free immunoassays. Conventional SPR is applied in specialized bio sensing instruments. These instruments use expensive sensor chips of limited reuse capacity and require complex chemistry for ligand or protein immobilization. Laboratory has successfully applied SPR with colloidal gold particles in buffered solutions. This application offers many advantages over conventional SPR. The support is cheap, easily synthesized, and can be coated with various proteins or protein ligand complexes by charge adsorption. With colloidal gold, the SPR phenomenon can be monitored in any UV spectrophotometer. This simple technology finds application in label free quantitative immunoassay techniques for proteins and small analytes, in conformational studies with proteins as well as real time association dissociation measurements of receptor ligand interactions for high throughput screening and lead optimization.

Evanescent wave sensors exploit the properties of light totally reflecting at an interface and the presence of an evanescent field of light at this interface. This technique makes use of the exponentially decaying electromagnetic field at the boundary between two media of different optical thickness upon irradiation with electromagnetic waves. Under total internal reflection conditions the decay length of the evanescent field into the optically thinner medium is of the order of the wavelength of the used excitation light. For visible light the field decays within a few hundred nanometers. Only analyte molecules in the evanescent region are probed, which causes the surface sensitive character of such methods. Basically, three different evanescent wave formats are known: planar waveguides, fiber-optics and surface plasmon resonance devices.

A waveguide consists of a planar glass surface with a refractive index higher than the adjacent medium. Under certain conditions light coupled into this waveguide can travel through the sample by total internal reflection. An evanescent field can interact with molecules in the region surrounding the waveguide. Adsorbed analytes change the optical properties of the waveguide and alter the boundary conditions for guiding light in the sample. Hence, the light coupling out of the waveguide can then used to monitor binding reactions at the surface of the waveguide. Fiber-optic sensors utilize the same principle as waveguides, but differ in the experimental geometry.

In surface plasmon technique, however, the evanescent light wave is used to excite the nearly free electron gas in a thin film (~50 nm) of metal at the interface. The excitation of these so called surface plasmons, are directly dependent on the optical properties of the adjacent medium where the deposition of an optical mass on the metal surface will lead to a change in the coupling conditions of the evanescent wave with the plasmons. The excitation of the resulting surface waves gives rise to a field enhancement compared to the intensity of the incident electromagnetic field. This is used to detect mass changes of the film and thus to measure binding processed at the interface. Illumination by laser light can be used to excite the plasmons in metals. Then the system responds to changes in the optical properties of the medium close to the metal film by altering the intensity of the reflected light. For surface sensitive investigation of adsorption and desorption processes on metallic substrates, surface plasmon resonance is the method of choice. Commercial instruments are available and are routinely used to measure bio-molecular interactions.

Biosensor development is an extremely active area of research generating large numbers of publications annually and a constant stream of innovative ideas for steadily pushing back detection limits. Micro/nanofabrication technologies developed originally for the telecommunication industry have allowed researchers and engineers to even smaller and more complex biosensing devices. In order to maintain this trend and guarantee the efficiency of biosensors or sensors more generally, it is worthwhile combining more than just one detection mechanism in a sensor device. An error in one measurement mechanism can be detected by the other which might be more sensitive in measuring another totally different parameter however related mathematically with one another.

In this way errors in one measurement mechanism can be detected and corrected in the other measurement mechanism.

1.3 Aim of this thesis:

The aforementioned technique of surface plasmons resonance using the evanescent wave field is not new. The selective process of an analyte binding onto a biological compatible surface is always accompanied by two processes namely: a change in the electron density of the immediate vicinity where the binding takes place and secondly a change in the effective layer thickness. This two processes can be monitored independent using x-ray spectroscopy or to be very precise x-ray reflectometry. Therefore while it is possible to follow reactions or dynamic processes using the phenomena of SPR which measures changes in the dielectric constant of the medium, the same process can be simultaneously followed in situ independently by x-ray reflectometry which measures not changes in the dielectric constants but changes in the electron density and thickness of the medium.

A combination of both techniques i.e. SPR and x-ray reflectometry, guarantees the independent reproducibility of either SPR responses using results obtain from x-ray reflectivity or vice-versa. In this way errors which may arise in one measurement technique due to device precisions etc. can be easily foretold by the other measurement technique independently. Thus more precise results of dynamic processes or analyte binding on the surface of ligands can be ascertained.

This thesis therefore combines the process of SPR and x-ray reflectivity in one device for the measurement of static as well as dynamic process tested at modelled systems.

In the first chapter of this work, emphasis will be laid on the equations of Maxwell as foundation to the theoretical description of SPR.

The concepts of evanescent wave fields and its quantum origin will be discussed and derived mathematically since it's very fundamental to the process of surface plasmons resonance. The method of x-ray reflectivity and the necessary parameters obtain from this method will be derived in chapter 2 of this work. Then the case will be made why a combination of both measurement methods can help enhance the general precision of measurements made by one method. The various experimental methods used in both the SPR as well as the x-ray reflectivity measurements will be explained in chapter 3.

In chapters 4 and 5 static and dynamic measurements combining both techniques will be discussed. The dynamic diffusion of gold colloids of varied circumference through poly (3-hexylthiophene) P3HT will be studied in detail using theoretical model as well as simulation program for both measurement techniques.

2 Theory

2.1 Evanescence Wave Optics

The term 'evanescence' is derived from the latin verb 'evernescere' which means to vanish. It summarises a number of optical phenomena related to electromagnetic waves decaying perpendicular to interfaces between two or more media with different optical properties. If for example a plane wave falls on the interface between a higher and lower refractive index media, from the higher refractive index side the wave will be totally internally reflected at the interface from a certain angle of incidence onwards the so called angle of total reflection θ_C. The rigorous theoretical treatment of this problem shows that the electric field along the propagation direction at the interface is still oscillating as usual while its component perpendicular to the interface is decaying exponentially. The decay depth is of the order of the wavelength used and is described by the following equation:

$$d = \frac{\lambda}{2\pi n_2} \left(\frac{n_1^2}{n_2^2} \sin^2 \theta_i - 1 \right)^{\frac{1}{2}} ; \theta_i > \theta_C \qquad 2.1$$

Here n_1 and n_2 denote the refractive indices of the two media ($n_1 > n_2$) and θ_i denotes the angle of incidence as measured from the normal to the surface. Such surface bound decaying waves are the basis for many experimental techniques that characterise surface properties. A change in the optical properties of the medium within the decay length of the field causes a response in the field properties. This response is exploited by various types of spectroscopy for example total internal reflection fluorescence spectroscopy [12]. In this thesis only an extension of the above phenomena is used. This is concern with surface plasmons paritons, the excitation of the loosely bounded or frees electrons between the interface of a noble metal and a dielectric. A major part of this work is concern with the study of the coupling capacity of the evanescence wave field while systematically varying the electron density of the medium.

In this respect, a mathematical description of electromagnetic waves and their interaction with interfaces will be elucidated, the case of one of the two media forming the interface being the metal is presented, and the two different ways of exciting surface plasmons namely prism and grating coupling are discussed with special emphasis to prism coupling, the method used throughout this work. Finally the Parrat formalism method which is able to describe the optical reflection and transmission and reflection behaviour of multiple layers will be discussed.

2.1.1 Plane wave at interfaces

Material properties and the geometry of the media influence the propagation of light through different materials. Mathematically, the description is based on Maxwell's equations which form the basis of classical electrodynamics. By virtue of the fact that evanescenct optical waves occur at the interfaces between two dielectric media, the most important processes at such an interface i.e., reflection, refraction and transmission are addressed.

2.1.1.1 Maxwell's Equations and Plane waves

With the simplification of isotropic, homogenous media electromagnetic radiation is described by Maxwell's equation without any source terms.

$$\nabla \bullet B = 0; \nabla \times E = -\frac{\partial B}{\partial t}$$
$$\nabla \bullet D = 0; \nabla \times H = -\frac{\partial D}{\partial t}$$
2.2

Here E is he electric field, B the magnetic induction, H the magnetic field, and D the electric displacement. The relationships between B and E on the one side and between D and H on the other side are given by:

$$D = \varepsilon.E$$
$$B = \mu.H$$
2.3

where the dielectric constant ε and the magnetic permeability μ are generally complex tensors of a second rank. However for isotropic material as considered here these tensors reduces to simple scalars $\varepsilon\varepsilon_o$ and $\mu\mu_o$ with ε_o being the permittivity of vacuum and μ_o the permeability of free space. Then the solution of Maxwell's equation as a function of time t at a point r are plane waves which are characterised by their electrical field amplitude E_o, their wave vector k and their angular velocity ω given by the following equation:

$$E(r,t) = E_o \exp[i(kr - \omega t)] \qquad 2.4$$

The above equation means that only the real part has a physical meaning and the orientation of E_o is orthogonal to k. For each pair of (k, ω), two mutually orthogonal electric field amplitudes exist spanning the plain to give all possible polarisations. Beside the electric field, also the magnetic field with the same mathematical notations contains the full information about the plane wave. Both representations may be transformed into one another by the use of

$$H(r,t) = \frac{1}{\mu\mu_o\omega} \bullet kxE(r,t)$$
$$E(r,t) = \frac{1}{\varepsilon\varepsilon_o\omega} \bullet kxH(r,t). \qquad 2.5$$

The dispersion relation relates the modulus |k| of the wave vector to a given angular frequency ω

$$\frac{\omega^2}{|k|^2} = \frac{1}{\mu\mu_o\varepsilon\varepsilon_o} = \frac{c^2}{n^2} \qquad 2.6$$

Here the refractive index n is defined as the ratio of the speed of light c in vacuum. Making the assumption of a non-magnetic material, the dispersion equation can be further simplified to:

$$|k| = \omega\sqrt{\varepsilon\varepsilon_o\mu_o} = k_o\sqrt{\varepsilon} = k_o n \qquad 2.7$$

2.1.1.2 Fresnel Equations

The diagram below summarises the events that can occur when a plane electromagnetic wave of fixed wave vector k_i and frequency ω falls onto a flat interface between two dielectrics and homogenous media of refractive indices n_1 and n_2. At the interface the incident wave is subject to reflection and transmission, of the wave being refracted. A fundamental characteristic of all these processes is the conservation of the wave vector $k_\parallel$, parallel to the interfaces. However the modulus and direction of the wave vector its self is changed.
For the reflected beam this means that the angle of incidence equal the angle of reflection.
For transmission this conservation is mathematically expressed in Snell's law as follows:

$$n_1 \sin\theta_i = n_2 \sin\theta_t \qquad 2.8$$

where θ_i and θ_t are the angles of incidence and transmission measured from the surface normal. The conservation of $k_\parallel$ is also the reason why an electromagnetic wave field in a multiple layer system can be decompose into two components in each layer: an upward and a downward propagating plane wave that completely describes the whole area.

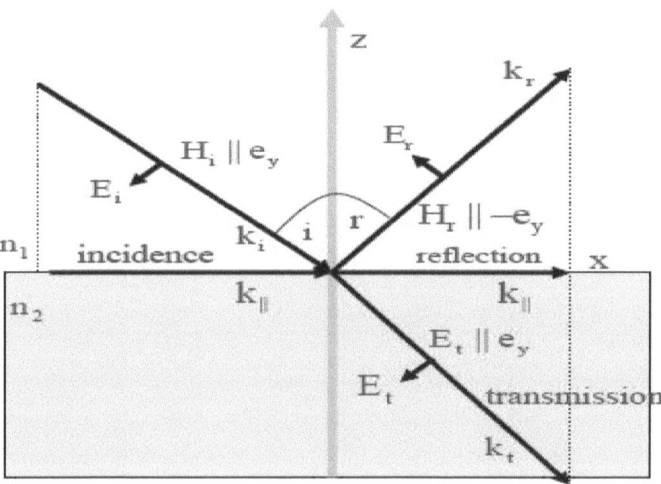

Figure 2.1: Schematic picture dielectric media. The plane of incidence is spanned between the wave vector of the incident light beam and the surface normal. In- plane wave vectors arte equal for all components of a reflected and transmitted transverse magnetic polarised plane wave at an interface between two different

The geometric construction of a planar surface leads to a straight forward choice of orthogonal polarisation vectors which are needed to fully describe a plane wave with (k, ω) and that don't change their state of polarisation upon reflection or transmission. On the one hand there is the transverse electric (TE) polarisation or s-polarisation having its electric field component perpendicular to the plane of incidence, which is spanned by the wave vector k_i and the normal to the surface. On the other hand we have the transverse magnetic (TM) mode or (p-polarisation) that possesses its only non-vanishing magnetic field component to the plane of incidence. In general, all polarisations may now be described by the linear superposition of the above mentioned two orthogonal polarisations. Usually in experiments, the wavelength of the light being used is known; the polarisation and the intensity of the incident beam are also known. If such a beam impinges on an interface to another medium, the analysis of the reflected or transmitted beam intensities gives inside into the properties of the material. Such reflectivity spectrum, i.e. intensity of the incident beam as a function of the angle of incidence, can be calculated by using the Fresnell's formulas. There are two different sets of formulas of reflection ratios r^{refl} and transmission ratios r^{trans} for the two different polarisations:

$$r_{s-pol}^{refl} = \frac{E_r}{E_i} = \frac{n_2 \cos\theta_i - n_1 \cos\theta_t}{n_2 \cos\theta_i + n_1 \cos\theta_t} \quad ; \quad r_{s-pol}^{trans} = \frac{E_r}{E_i} = \frac{2n_1 \cos\theta_i}{n_2 \cos\theta_i + n_1 \cos\theta_t}$$

$$r_{p-pol}^{refl} = \frac{E_r}{E_i} = \frac{n_2 \cos\theta_t - n_1 \cos\theta_i}{n_2 \cos\theta_t + n_1 \cos\theta_i} \quad ; \quad r_{p-pol}^{trans} = \frac{E_r}{E_i} = \frac{2n_1 \cos\theta_i}{n_2 \cos\theta_t + n_1 \cos\theta_i}$$

2.9

However, since in an experiment only intensities are detected, it implies that the ratio of the reflectivity R and the transmission T can be written as

$$R = \frac{I_r}{I_i} = \frac{|E_r|^2}{|E_i|^2} = (r^{refl})^2 \qquad T = \frac{I_t}{I_i} = \frac{|E_t|^2}{|E_i|^2} = (r^{trans})^2 .$$

2.10

The deduction of the Fresnel equation follows the route of the first setting up the appropriate plane wave equation for the incident, reflected und transmitted beams and then demanding that at z=0 the fields (and therefore exponentials) have to be equal. This result is the equality of the angles of incidence and reflection and Snell's law. The requirement that the tangential components of E and H have to be continuous leads then to the Fresnel's equations themselves.

2.1.1.3 Total Internal Reflection

The situation of light passing through medium 1 with the refractive index n_1 which is then reflected at medium 2 with refractive index n_2 that is smaller than n_1 gives rise to a special feature: Beginning at an angel of incidence θ_1, the transmission angle can be determined according to Snell's law. The increase of θ_1 leads to an increase of θ_2 up to a point where θ_2 reaches a value of 90°. The so-called critical angel θ_C is reached. At that point the reflectivity reaches a value of 1, i.e. light is reflected and any further increase of θ_1 has no influence on the reflectivity anymore. However at such high angles, the component of the field normal to the surface is no longer oscillatory but decaying exponentially as given by equation 1. This is the regime of evanescence waves.

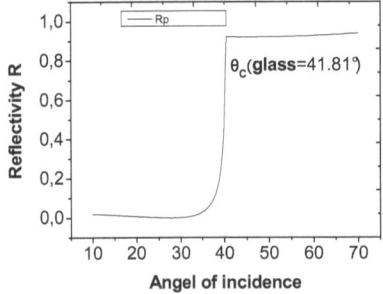 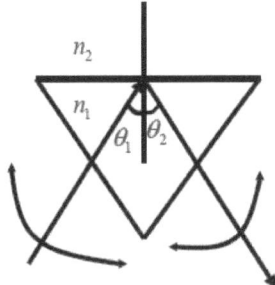

Figure 2.2: Simulated reflectivity as a function of the angel of angel of incidence for the system glass ($n_1 = 1.5$)/ air ($n_2 = 1$). Depicted is the p-polarisation at $\lambda = 780$nm and the corresponding reflection scheme.

2.1.2 Surface Plasmons Polaritons

Before moving on to the theoretical description of surface plasmons polaritons or surface plasmons for short, a brief descriptive understanding of surface plasmons is given. The excitation methods of surface plasmons is then presented with emphasis on prism coupling; the method used in our experiment. The grating coupling method will also be discussed. Finally focus is drawn onto how the system responds if extra layers are added to the dielectric. Wave-like electromagnetic modes that propagate along an interface between two media and whose amplitudes decreases exponentially normal to the surface are called surface polaritons, i.e. surface electromagnetic modes involving photons coupled to surface electric dipoles and/or magnetic dipole excitation [13]. A plane wave of transverse electric–dipole excitation propagating along the x-axis in an optically isotropic medium is now considered as shown in figure 2.3. Since the macroscopic polarisation P is transverse and $\nabla \bullet P = 0$ there are no volume polarisation charges and, thus no electric field exists. Now a non-dispersive dielectric medium is with a surface normal parallel to P. As a result of the discontinuity in P, a periodic surface charge density is established at the surface giving an electrical field with components along x and z. Due to the fact that the surface charge density alternates in signs, the magnitude of the field decreases exponentially in the direction normal to the surface. Furthermore, the surface charge density is the only source of the electric field and thus its z-component at equidistant points from the interface are opposite in sign. However, since the normal component of the electric displacement D at the interface on both media have to be continuous, it follows that the dielectric constants $\varepsilon_1(\omega)$ and $\varepsilon_2(\omega)$ have opposite signs. This is the basic condition for the existence of surface electric dipole excitations. When such an electrostatic field is coupled to 'surface photons' a so called surface polariton is created, the total electric field of which consists of a superposition of the constituting electrostatic and electromagnetic fields. Since the coupling photon has to provide for the surface charge density the surface polaritons are transverse magnetic (TM) modes.

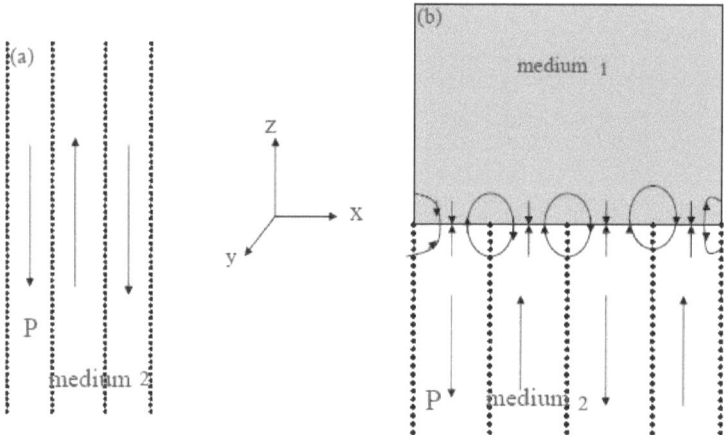

Figure 2.3 Schematic diagram of (a) a plane wave polarised along y and propagating in medium 2 along the x- axis. (b) Shows the surface charge density when a new medium 1 is introduced in the x-y plane.

2.1.2.1 Solving the Maxwell's Equations
Again the interface between two media of different frequency dependent, but this time complex dielectric functions

$$\varepsilon_1 = \varepsilon_1' + i\varepsilon_1''$$
$$\varepsilon_2 = \varepsilon_2' + i\varepsilon_2''$$

2.11

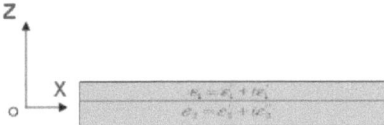

Figure 2.4: Schematic representation of 2 dielectricum with complex dielectric functions.

is examined. The linked between the complex dielectric constant ε and the complex refractive index $(n+i\kappa)$ is given by:

$$(n+i\kappa)^2 = \varepsilon' + i\varepsilon'' = \varepsilon$$
$$\varepsilon' = n^2 - \kappa^2 \qquad 2.12$$
$$\varepsilon'' = 2n\kappa$$

The real part n is called the refractive index while the imaginary part κ is the absorption coefficient. It is responsible for the attenuation of an electromagnetic wave. The magnetic permeability μ_1 and μ_2 are considered to be unity.

As earlier explained, there only exist surface polaritons for transverse magnetic polarised incident plane waves. Thus the solution to the problem will take the following form:

$$A_1 = A_{01} \exp[i(k_{x1}x + k_{z1}z - \omega t)]$$
$$A_2 = A_{02} \exp[i(k_{x2}x + k_{z2}z - \omega t)] \qquad 2.13$$

For A either being the magnetic H or the electric field E. k_{x1} and k_{x2} are the wave vectors in the x-direction while k_{z1} and k_{z2} are the wave vectors along the z-axis. The indices 1 and 2 are reference to the two media involved for z>0 and z<0 respectively. The continuity of the tangential component of E and H at the surface, i.e. $E_{x1} = E_{x2}$ and $H_{y1} = H_{y2}$, leads to:

$$k_{z1}H_{y1} = \frac{\omega}{c}\varepsilon_1 E_{x1}$$
$$k_{z2}H_{y2} = \frac{\omega}{c}\varepsilon_2 E_{x2} \qquad 2.14$$

On the other hand inserting equation (2.13) into Maxwell's equations (2.2) gives

$$k_{x1} = k_{x2} = k_x. \qquad 2.15$$

This leads to the only non-trivial solution if

$$\frac{k_{z1}}{k_{z2}} = -\frac{\varepsilon_1}{\varepsilon_2}. \qquad 2.16$$

For gold, the dielectric constant at a wavelength of 780nm is -22,4+i1,4 and P3HT with no absorption component of its dielectric constant (no complex component) has a dielectric constant of 1,06. Hence for a P3HT/Gold interface $\frac{k_{z1}}{k_{z2}}$ >0, hence it is define.

In words this means that surface electromagnetic modes can only be excited at such interfaces where both media have dielectric constant of opposite signs, as has already been shown above. If one of the two media is a dielectric with a positive dielectric constant ε_d then the above relation can be fulfilled by a whole variety of possible elementary excitations if and only if their oscillation strength is strong enough to result in a negative dielectric constant ε. For excitations like phonons or exitons, the coupling to a surface electromagnetic wave leads to a phonon polariton or exiton surface polariton modes, respectively. Another form of excitation that can couple to surface electromagnetic waves is the collective plasma oscillation of a nearly free electron gas around the charged metal ions, called surface plasmons polaritons. In dielectrics, the electrons are bounded tightly to the nuclei resulting to a small, positive and real dielectric constant. In metals however the electrons are quasi-free and may be moved easily by an external force. The classical Drude model [60], which considers the electron to be free, already derives a highly negative, complex dielectric constant:

$$\varepsilon(\omega) = 1 - \frac{\omega_p^2}{\omega^2} \qquad 2.17$$

The plasma frequency ω_p always lies in the visible range for metals. The above equation is valid for frequencies ω from 0 up to a maximum frequency ω_{max}, which is given by

$$\omega_{max} = \frac{\omega_p}{\sqrt{1+\varepsilon_d}}. \qquad 2.18$$

For metals the dielectric function, ε_m, is in general complex with a negative real-part and a small positive imaginary part. The complex component is responsible for absorption and hence explains the decay of electromagnetic waves in metals. The negative part of the real component guarantees for a positive wave number.

Continuing the above deduction of the very distinct wavevector of a surface plasmon, the wavevectors in the direction of the z-axis can be calculated from equations (2.1) and (2.14):

$$k_{zd} = \sqrt{\varepsilon_d \left(\frac{\omega}{c}\right)^2 - k_x^2}$$

$$k_{zm} = \sqrt{\varepsilon_m \left(\frac{\omega}{c}\right)^2 - k_x^2}.$$

2.19

Finally deploying equation (2.16), we obtain the dispersion equation for surface plasmons at a metal/dielectric interface:

$$k_x = k_x' + ik_x'' = \frac{\omega}{c}\sqrt{\frac{\varepsilon_m \cdot \varepsilon_d}{\varepsilon_m + \varepsilon_d}}.$$

2.20

The complex nature of the wave vectors in the x- and z-direction leads to an exponentially decaying wave in the z-direction and an attenuated wave along the x-axis. A finite propagation length L_x where

$$L_x = 1/k_x''$$

2.21

can be defined. This quantity extremely influences the lateral resolution and is especially important in surface plasmons spectroscopy applications. For a gold/air interface with $\varepsilon_m = -22.4 + i1.4$ at $\lambda = 780$nm, the propagation length is in the range of 10μm. The penetration depth δ_d and δ_m into the dielectric and metal respectively are both defined as:

$$\delta_d = 1/k_0 \left|\frac{\varepsilon_m' + \varepsilon_d}{\varepsilon_d^2}\right|^{\frac{1}{2}}$$

$$\delta_m = 1/k_0 \left|\frac{\varepsilon_m' + \varepsilon_d}{\varepsilon_m'^2}\right|^{\frac{1}{2}}$$

2.22

These are the length scale by which the intensity falls to 37% of its initial value into the dielectric and metal layer.

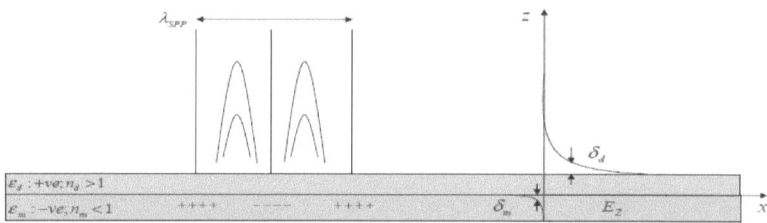

Figure 2.5: Schematic representation of the penetration depth of evanescence wave into the metal and dielectric.

2.1.2.2 Excitation of Surface Plasmons

Another aspect of the dispersion relation of surface plasmons (2.20) is summarised in the following equation:

$$k_{x,SP} = \frac{\omega}{c}\sqrt{\frac{\varepsilon_m \varepsilon_d}{\varepsilon_m + \varepsilon_d}} \geq \frac{\omega}{c}\sqrt{\varepsilon_d} = k_{x(max),ph} \qquad 2.23$$

One result of this equation is, as already stated above, that the z-component of the surface plasmons wave vector is imaginary. Therefore the surface plasmon wave is a non-radiative evanescence wave with its maximum field amplitude at the interface. It decays exponentially into the dielectric and metal. Another consequence is that a light beam incident from the dielectric with the maximum wave vector $k_{x(max),ph}$ at the interface (grazing incidence, i.e. the angel of incidence is 90° and therefore $\sin\theta_i = \sin 90° = 1$) cannot excite a surface plasmon with the wave vector $k_{x,SP}$ since its momentum is not sufficiently large.

These details are illustrated graphically in more detail in figure 2.6. Although the light line of the photons (a) approaches asymptotically the dispersion curve of surface plasmons (SP1), there is no intersection of both curves and the x-component of the of the wave vector of incident light is always smaller than the one for surface plasmons. Methods to increase the momentum of the light in order to couple to surface plasmons include nonlinear coupling [13] or coupling by means of a rough surface [11]. The most predominant by far however are the prism coupling and the grating coupling technique. The prism coupling technique will be discussed in detail while the grating coupling method will be mentioned.

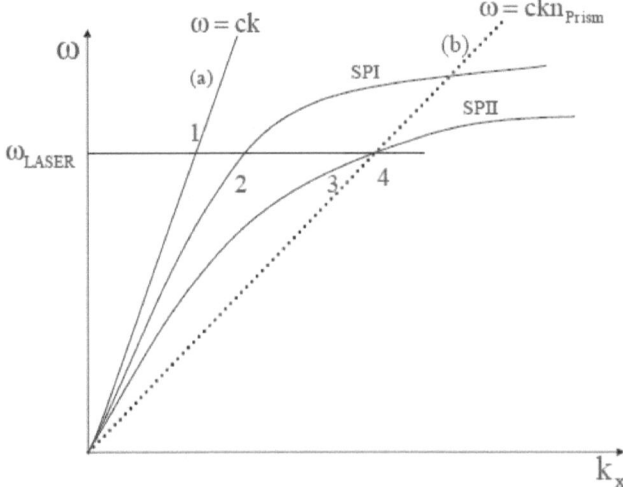

Figure 2.6: Dispersion relations of free photons in a dielectric (a) and in a coupling prism (b) with $n_p > n_d$, compared the dispersion relation of surface plasmons at the interface between metal and dielectric before (SPI) and after (SPII) the absorption of an additional dielectric layer. At a given laser wavelength ω_{Laser} the energy and momentum match of the photons impinging from a dielectric with the surface plasmons is not achieved (1) were as for photons incident through a prism this is attained. (2, 3)

2.1.2.2.1 Prism Coupling

As already pointed out earlier, prism coupling represents one of the ways of increasing the wave vector of the incident light and therefore the x-component of the wave vector, which only couples to the surface excitations. Figure 4 shows the corresponding dispersion relation if the refractive index of the prism n_p is larger than the one of the dielectric n_d. The momentum is increased, the curve is more tilted to the right and therefore at a given laser wavelength ω_{Laser}, coupling to surface plasmons (2, 3) can be obtained. However since at point 4 of fig. 2.6 the momentum of the light beam is too large, it has to be tuned to the one of the surface plasmons by varying the angle of incidence ($k_{x,Ph} = |k_{ph}| \sin \theta_i$).

There exist two different configurations with which to excite surface plasmons by use of a high refractive index prism. The one which was proposed first is the so-called Otto configuration [17]. Here the laser beam is reflected off the base of a prism (common geometries are half spheres, half cylinders or 90° prism). The evanescence radiation couples from the totally internally reflecting base of the prism to the bound surface field of the surface plasmon. Experimentally, the resonant coupling is observed by monitoring the reflected light beam as a function of the incident angle. However there is a major technical drawback to this type of configuration as one has to fulfil the need of providing gab of approximately 200 nm for effective coupling. Even a few dust particles can act as spacers preventing a controlled assembly of the coupling system.

The other method of coupling light to plasmons by means of a high refractive index prism is known as the Kretschmann configuration [10; 35]. In this excitation scheme, the light does not couple through a dielectric layer yet, alternatively through a thin metal layer which is directly everporated onto the base of the prism. At the momentum matching condition, a surface plasmon is then excited at the interface between metal and dielectric as depicted in figure 5.

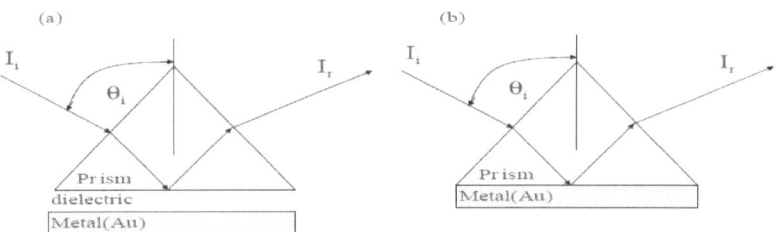

Figure 2.7: Coupling geometries for high refractive index prism: (a) shows the Otto configuiration and (b) the Kretschmann configuration. In both case the surface plasmon propagates along the metal/dielectric interface. Coupling is only possible if the refractive index of the prism is higher than that of the dielectric.

However in contrast to the above derived mathematical description, the surface plasmons are not restricted to two-half spaces any more. Quantitatively, one has to take the thickness of the metal layer into account, which allows in particular that some of the surface plasmons light are coupled out through the metal and the prism. This new additional radiative loose channel however can be considered a minor disturbance to the surface plasmons electromagnetic wave

[18, 19]. In any case, it is clear that there exist an optimal thickness of the metal. Assuming that the metal film is too thin additional damping of the surface plasmon waves will occur due to the radiative loose channel back through the metal film and the prism. If the metal layer is too thick the tunnel barrier is too thick and too large and only little light will couple to surface plasmons at the metal/dielectric interface. For gold the optimal thickness for a laser with wavelength $\lambda = 780$nm lies between 45 and 60 nm. This optimal thickness was investigated by considering the full width at half maxima and the dip of the plasmon coupling curves for different thicknesses of gold.

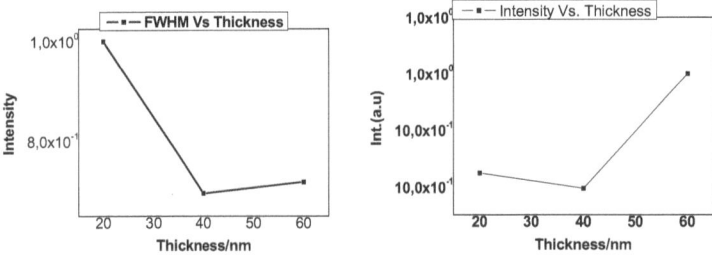

Figure 2.8: Variation of the FWHM and coupling intensity as a function of the thickness of the gold layer of different thickness.

This thickness can be controlled by evaporation rate of gold.

2.1.2.2.2 Bare Metal Reference

As high refractive index prisms are used for the excitation of surface plasmons in the next examples, the momentum of the incident light beam in the plane of the interface exceeds the one needed for the excitation. Thus it is possible to tune the system into resonance by simply changing the angle of incidence, since $k_x = k_o \sin \theta_i$. At lower angles, the reflected intensities increases as described by the formulas of Fresnell. Then from a certain angle, the angle of total internal reflection θ_C onwards, it reaches a maximum. Note that the reflectivity before θ_C is rather high compared to the type of interface shown on figure 2.2. This is due to the evaporated metal film that acts as a mirror reflecting most of the incident light. Secondly the maximum reflected intensity never reaches unity since the photon energy is partly dissipated on the metal layer. Lastly the position of the critical angle only depends on the substrate and superstrate, i.e. prism and water. If the projection of k_i to the interface matches $k_{x,SP}$, resonance occurs and a surface plasmon is excited. This condition is given at the intersection (2) of figure 2.6. At resonance, the electromagnetic waves couples to the surface plasmon. As a consequence of this coupling the measured out going intensity is at a minimum. The minimum is denoted by θ_0 and is given by:

$$\theta_0' = \arcsin\left(\frac{1}{n_p}\sqrt{\frac{\varepsilon_m \varepsilon_d}{\varepsilon_m + \varepsilon_d}}\right) \qquad 2.29$$

with n_p being the refractive index of the coupling prism.

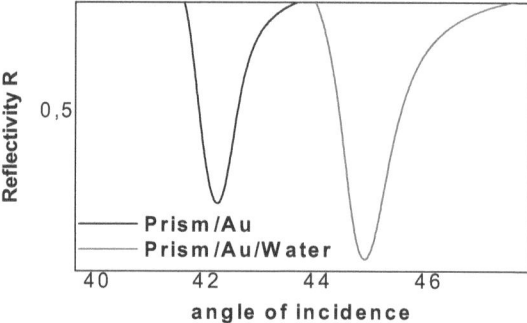

Figure 2.10: Simulated reflectivity as a function of angle of incidence for the system prism/gold/water. As basis for calculations, the following parameters are used: $n_p = 1.55$; $\varepsilon_{Au} = -22,4 + i1.4$, $n_{water} = 1.332$, the calculated resonance angles are: prism/gold = 40,7°; Prism/gold/water = 45,2°.

As mentioned already, for real metals there is resistive scattering and hence damping of the oscillations created by the incident electric field. If this was not the case, the surface plasmon would be infinitely sharp with an infinite propagation length. The imaginary part of the dielectric constant of the metal causes the damping. With this consideration the dispersion relation for surface plasmons can be re-written as follows:

$$k_x = k_x' + ik_x'' = \frac{\omega}{c}\sqrt{\frac{(\varepsilon_m' + i\varepsilon_m'').\varepsilon_d}{(\varepsilon_m' + i\varepsilon_m'') + \varepsilon_d}} \approx \frac{\omega}{c}\sqrt{\varepsilon_d}\left(1 - \frac{\varepsilon_d}{2\varepsilon_m'}\right) + i\frac{1}{2}\frac{\omega}{c}\frac{\varepsilon_m''(\varepsilon_d)^{\frac{3}{2}}}{(\varepsilon_m')^2} \qquad 2.30$$

Thus the shift of the resonance peak is inversely proportional to ε_m' whereas the width, which is related to k_x'' depends on ε_m'' and is inversely proportional to $(\varepsilon_m')^2$. While at first sight it might be advantageous to have a small imaginary part of the metal dielectric, the real part is of even higher significance. An example of this finding is given in figure 2.11 where surface plasmons resonance curves for gold and silver are compared to each other. $\varepsilon_{Au} = -22.4 + i1.4$ and $\varepsilon_{Ag} = -17 + i0.7$. Clearly silver with a higher absolute value of ε_m' and the smaller imaginary part can be identified having a much sharper resonance.

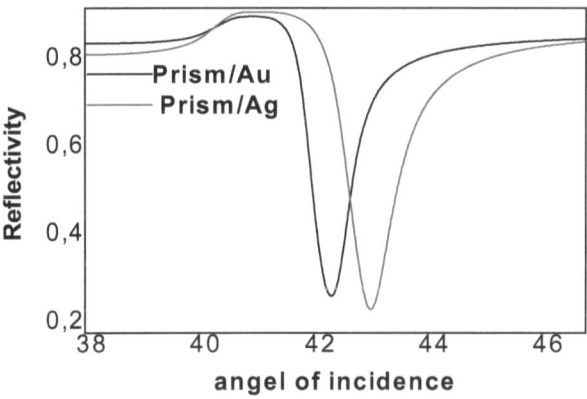

Figure 2.11: Variation in resonance for Au and Ag due to differeces in ε_m' and ε_m'' for both metals, Resonance angles: Prism/Au= 42, 18°; Prism/Ag= 42, 87°.

It must be mentioned that there does exist a difference between the shape of surface plasmons exited at a plane interface and at a surface grating. For the prism coupling the actual position of the surface plasmons resonance depends on the refractive index and consequently the type of glass of the prism. The slope of the dispersion curve of light depends inversely on the square root of the dielectric constant of the material. A higher refractive index prism consequently shifts the resonance curve to lower angles of incidence. The shape of the resonance curve of gratings is calculated using Fresnel's formulas. On the contrary, the position and shape of grating coupled surface plasmons depends on the shape of the surface relief. Hence it varies from grating to grating. The position of the edge of the total reflection is a function of wavelength of the grating and the coupling strength is dictated by the groove depth, since this reflects the clarity of the edge of total reflection. This last parameter though cannot be easily controlled as the evaporation thickness.

2.1.2.2.3 Thin Additional Films

The major advantage of surface plasmons lies in its sensitivity to surface processes due to its evernescence wave field. This means, on the one hand, that a change of the substrate leads to either an increase or a drop of the wavevector of the surface plasmons resonance. On the other hand, the addition of a thin layer with a thickness up to d where $d << 2\pi/k_{z,d}$ of a second dielectric to the already existing layer triggers a changed surface plasmons response and the corresponding shift of the dispersion curve is equivalent to a change of the overall refractive index integrated over the evernescence field. The net effect is a shift in the surface plasmons curve as can be seen in figure 2.10 for an additional layer with higher refractive index than the one of the reference dielectric. At the same energy $\hbar\omega$ for incident light, the dispersion curve of the surface plasmons intersects with the light line at higher wave vectors (point 4 of figure 2.6). In terms of the reflectivity as a function of the angle of incidence, the minimum is therefore shifted to higher angles.

When adding a layer to an existing system, two parameters are of interest namely the refractive index and the thickness of the film. In order to separate these two parameters at least two distinct features that are correlated to the addition are needed. Yet surface plasmons resonance only provides one. Consequently only a set of parameters (n, d) can be derived from such reflectivity curves, provided both parameters are unknown. If one of them is known, the other can be obtained by fits on the curves. Several methods resolve the ambiguity of this problem: first resonance curve can be taken at different laser wavelengths [22]. This method however does not resolve the ambiguity of the unknown dispersion behaviour of the refractive index of the coating. Secondly the contrast of the experiment can be varied [23] i.e. the surface plasmons curves are measured in at least two solvents with varied refractive indices. The minimum shift does not depend on the absolute value of n but on the contrast i.e. the refractive index difference between the layer and the surrounding medium. In both methods, a set of at least two different curves of n vs. d is obtained, the intersection of which determines the correct refractive index and the thickness of the additional layer.

3 X-Ray Reflectivity

We have seen that the addition of a layer on the surface of a metal supporting surface plasmons resonance leads to a shift of the resonance angle either to the left or to the right depending on how the evernescence wave field couples with the electron density of the added layer. We have also seen that absolute determination of the effective thickness of the newly introduced layer is not directly possible using just one set of measurements from the surface plasmons resonance stand point.

However the effective determination of layer thickness and its associated parameters is very vital in modern technological devices. This is because most properties of thin films are thickness dependent for example the drain current of field effect transistors is strictly dependent on the thickness of the body carrying the charge carriers. Therefore determination of film thickness is very crucial for modern technology. X-ray reflectometry is a non-destructive and non- contact technique used in determining thicknesses of films between 2-200 nm with a precision between 1-3 Å. In addition to thickness determination the technique is also used to determine density and roughness of films with extremely high precisions. It is in the light of these new features associated with this technique, and the fact that x-ray reflectometry determines the electron density profile of thin layers which made it interesting to deploy these two independent techniques in a single device. While it is possible to determine with high precision the electron density profile of an additional layer using x-rays, the technique of surface plasmons resonance can independently determine the dielectric constant of the thin layer. From the relationship [29]:

$$\varepsilon_{LD}(r, r', \omega) = \left(1 - \frac{4\pi e^2 n(r')}{m\omega^2}\right) \delta(r - r'), \qquad 3.1$$

it's obvious that a model explaining the trends in the electron density distribution of the thin film can be used to calculate with extremely high precision the trends in the plamons response independently. It is in this light that this research makes use of both techniques simultaneously. In the subsequent sections the basic principles of x-ray reflectivity will be elucidated, followed by reflected intensities from ideally flat surface. The last sub-section will be dedicated to the acquisition of data from x-ray experiments in our home laboratory and its subsequent analysis and simulations.

3.1 Basic Principles of X-Ray Reflectivity

X-rays are part of the broad spectrum of electromagnetic waves. They can be produced by the acceleration or deceleration of electrons either in a vacuum in which case we talk of synchrotron radiation or in metallic target called tubes [24]. The most widely used x-rays in material science have a wavelength of the order of 0.1nm. This wavelength is associated with a very high frequency of 1019 Hz which is at least four orders of magnitude greater than the eigen frequency of an electron bounded to the nucleus. As a consequence, the interaction of x-rays with matter can be well described (in a classical way for a first approach) by an index of refraction which characterises the change of direction of the x-ray beam when passing from air to a material. A very simple classical model in which an electron of the material is considered to be accelerated by the x-ray field shows that the index of refraction of x-rays can be written as $1-\delta-i\beta$ where δ and β account for the scattering and adsorption respectively. The sign preceding β depends on the convention of signs used to define the propagation of the electric field. The values of δ and β (which are positive) are related to the electron density and linear absorption coefficient of the material through the following relations:

$$n = 1 - \sum_k \frac{Z_k + f_k' + if_k''}{V_m} \quad , \qquad 3.2$$

where

$$\delta = \frac{r_e}{2\pi}\lambda^2 \sum_k \frac{(Z_k + f_k')}{V_m} = \frac{r_e}{2\pi}\lambda^2 \rho , \qquad 3.3$$

and

$$\beta = \frac{r_e}{2\pi}\lambda^2 \sum_k \frac{f_k''}{V_m} = \frac{\lambda}{4\pi}\mu . \qquad 3.4$$

Here $r_e = 2.818 \times 10^{-6}$ nm is the classical radius of an electron, V_m is the volume of the unit cell Z_k is the number of electrons of atom k in the unit cell, f' and f'' are the real and imaginary parts of the absorption for the specific energy of the incident radiation. The sum is performed over all the atoms of the unit cell.

3.2 The critical angel of reflection

For x-rays the refractive index of a material is slightly less than unity. Therefore passing from air (n=1) to the refracting material (n>1), it is possible to totally reflect the beam if the angel of incidence α (angel between surface of sample and incident beam) is small enough. This is known as total external reflection of x-rays. For this to occur, the incident angle must be smaller than the critical angle α_C defined as $\cos(\alpha_C) = n = 1 - \delta$. Since n is close to 1, this angle is very small hence a Taylor approximation in α_C yields the following relation:

$$\theta_C^2 = 2\delta = \frac{r_e \lambda^2}{\pi}\rho \qquad 3.5$$

3.3 Reflected intensity from ideally flat surfaces

When an x-ray beam impinges on a flat surface, part of the incoming wave is reflected and part is transmitted through the material. If the surface of the reflecting material is flat, the reflected intensity will be confined in a direction symmetric from the incident one and will be labelled as specular. The specular reflectivity is conventionally defined as the ratio

$$R(\theta) = \frac{I(\theta)}{I_o}. \qquad 3.6$$

Here $I(\theta)$ is the reflected intensity at an angel θ and I_o is the intensity of the incident beam. The domain of validity of the x-ray reflectivity is limited to small angles of incidence where it is possible to consider the electron density as continuous. In this approximation, the reflection can be treated as classical problem of reflection of an electromagnetic wave at an interface. This leads to the classical Fresnel relationship, which as seen earlier, gives the reflection coefficient for s and p polarisation. The reflectivity which is the modulus square of this coefficient can be formulated in the case of x-rays as follows:

$$R^{flat}(\theta) = rr^* = \left|\frac{\theta - \sqrt{\theta^2 - \theta_C^2 - 2i\beta}}{\theta + \sqrt{\theta^2 - \theta_C^2 - 2i\beta}}\right|^2. \qquad 3.7$$

This expression is independent of the polarisation. Since the reflectivity is only observed in specular conditions (angle of incident equals angle of reflection), we obtain after introduction of the wave vector transfer $q = (0, 0, q_z = 4\pi \sin\theta/\lambda)$,

$$R^{flat}(q_z) = \left| \frac{q_z - \sqrt{q_z^2 - q_C^2 - \frac{32i\pi^2\beta}{\lambda^2}}}{q_z + \sqrt{q_z^2 - q_C^2 - \frac{32i\pi^2\beta}{\lambda^2}}} \right|^2 . \qquad 3.8$$

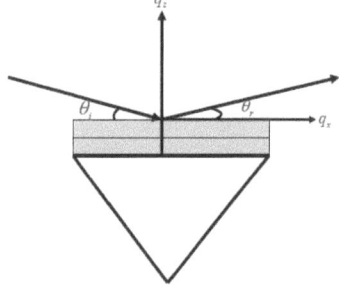

Figure 3.1: Schematic representation of x-ray reflectivity.

One recent tendency in present material research is the increasing ability to structure solids in one, two, and three dimensions to explore quantum effects at the sub-micrometer scale. Based on various material systems, artificially mesoscopic-layered superstructure such as multi layers, super lattices, layered gratings, quantum wires and dots have been fabricated successfully. This has opened new perspectives for manifold technological applications (e.g. anticorrosion coating, micro and opto-electric devices, magneto optical recording etc etc). This perfection of mesoscopic layered structure is characterised by:

1. The perfection of superstructure (grating shape, periodicity, layer thickness etc.
2. The interface quality (roughness, interdiffusion etc.
3. Crystalline properties (strain, defects mosaicity, etc)

X-ray specular reflectometry is used to measure the thickness of individual layers, the vertical spacing of the multiple layer stacking, the surface and interface roughness and the average density of a layered system.

These properties cannot be determined using results from surface plasmons resonance experiments alone.

From the laws of Fresnel's reflectivity, the intensity leaving a smooth surface decreases rapidly as the angel of incidence increases. To record intensities over more than 6 orders of magnitude, one needs a highly intensive incident beam and a detector with extremely low noise.

The main feature of the specular scans of a periodic multilayer is the multilayer Bragg peaks giving evidence for the vertical periodicity.

From general dynamic formula we can derive the expression for the reflectivity of a single layer deposited on a semi infinite substrate:

$$R = \left| \frac{r_1 + r_2 e^{-2ikT}}{1 + r_1 r_2 e^{-2ikT}} \right|^2 \qquad 3.9$$

Where $r_{1,2}$ are the Fresnel reflectivity coefficients of the free surface and the substrate interface, respectively k is the vertical component of the wave vector of the beam transmitted through the layer and T is the layer thickness.

For a general n layer system illustrated by the figure below, one can derive the following relations.

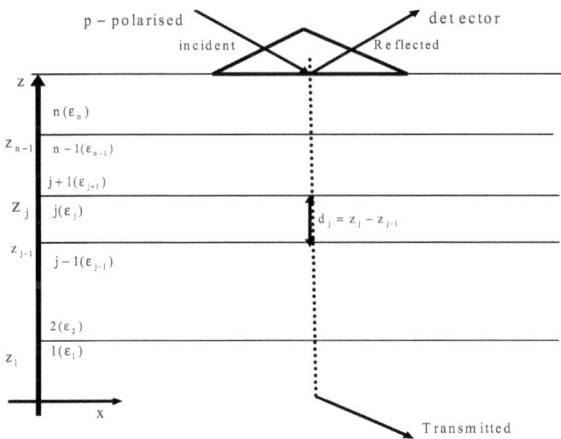

Figure 3.2: A layer model for the calculation of Fresnel reflectivity.

Starting again with a single layer system, we can represent the reflected beam from a stack of layers (the semi-infinite media ε_0 and ε_f, and the layer ε_1) as a superposition of waves:

a) reflecting from the interface $\varepsilon_0 / \varepsilon_1$;

b) transmitting through the interface $\varepsilon_0 / \varepsilon_1$, then passing through the layer reflecting from the interface $\varepsilon_1 / \varepsilon_f$, passing through the layer again, and leaving the layer through the interface $\varepsilon_0 / \varepsilon_1$;

c) the wave penetrating in the layer, reflecting from the interface $\varepsilon_1 / \varepsilon_f$ twice, and reflecting from the interface $\varepsilon_0 / \varepsilon_1$ ones, passing the layer forward and back twice and leaving the layer through the interface $\varepsilon_0 / \varepsilon_1$, etc

The figure below represents such a one layer system

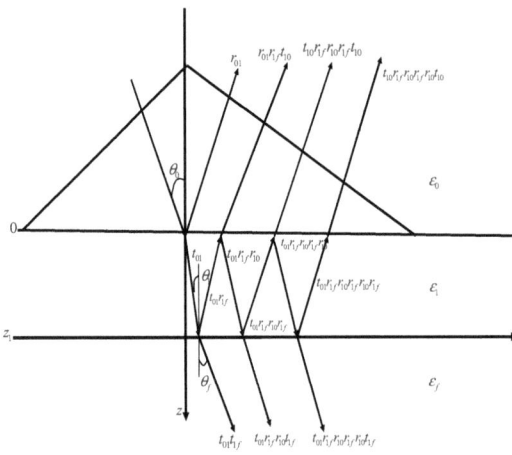

Figure 3.3: Application of multiple reflections to one layer.

Summing the contribution of all these partial waves to the total reflection coefficient R_1 we obtain:

$$R_1 = r_{01} + t_{01} \exp(ik_{1z}d_1) r_{1f} \exp(ik_{1z}d_1) t_{10} + t_{01} \exp(ik_{1z}d_1)$$
$$x[r_{1f} \exp(ik_{1z}d_1) r_{10} \exp(ik_{1z}d_1) r_{1f} \exp(ik_{1z}d_1) t_{10} \ldots\ldots\ldots$$

3.10

Beginning with the second term, we have an infinite geometric series. Summing it up and taking into account the fact that for any two successive layers j and j+1,

$$r_{j,j+1} = -r_{j+1,j}$$
$$r_{j,j+1}^2 + t_{j,j+1} t_{j+1,j} = 1$$

3.11

we obtain:

$$R_1 = r_{01} + \frac{t_{01} t_{10} r_{1f} \exp(ik_{1z} d_1)}{1 + r_{01} r_{1f} \exp(ik_{1z} d_1)} = \frac{r_{01} + r_{1f} \exp(ik_{1z} d_1)}{1 + r_{01} r_{1f} \exp(ik_{1z} d_1)}$$

3.12

For a two layer system, the picture of partial reflected and transmitted waves is so complicated that there is no direct and exact method to sum all partial waves contributing to R_2. However a recursive calculation of the reflection coefficient for a double layer system as shown below yields the following expression:

$$R_2 = \frac{r_{01} + r_{12} \exp(ik_{1z} d_1) + r_{2f} \exp(ik_{1z} d_1) \exp(ik_{2z} d_2) r_{01} r_{12} r_{2f} \exp(ik_{2z} d_2)}{1 + r_{01} r_{12} \exp(ik_{1z} d_1) + r_{12} r_{2f} \exp(ik_{2z} d_2) r_{01} r_{2f} \exp(ik_{1z} d_1) \exp(ik_{2z} d_2)}$$

3.13

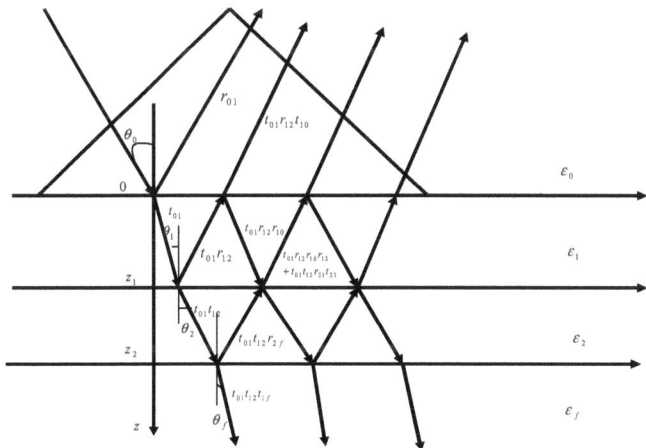

Figure 3.4: Application of multiple reflection method to two layers.

From the surface plasmons resonance perspective, the reflected coefficients $|R_1|^2$ and $|R_2|^2$ are only dependent on the thickness of the layer as shown by the simulation results below between polymer layers of different thickness on a gold layer of fixed thickness.

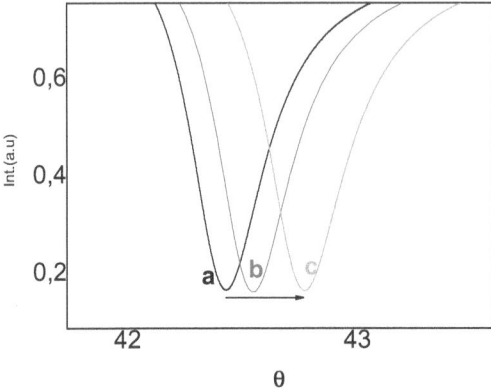

Figure 3.5: Thickness variation effect on reflected intensity. (a) Prism/gold 60 nm/ P3HT 45 nm, (b) Prism/gold 60 nm/ P3HT 60 nm, (c) Prism/gold 60 nm/ P3HT 90 nm.

Using the relationship:

$$\Delta\theta = (n*d)_{P3HT} + c \qquad 3.14$$

where (n*d) is the optical thickness of the layer and c an arbitrary constant, the variation between θ_0^P, R_{min} and d can be easily established.

This formalism treats the formed thin film as infinite, continuous dielectric layer. The analysis of the reflected intensity plots usually requires numerical calculations where the unknown parameters are obtained by fitting calculated results to experimental plots [25; 26].

In a dynamic process however, such as the thermal diffusion of gold colloids through a layer of polymer, it is often difficult to resolve how the different variables of the experimental system govern the different features such as θ_0^P and R_{min}, of the observed plots since the exact chemical architecture of the layered system and its constitution after each annealing process is not exactly known. To circumvent this shortcoming, an independent experiment which gives insight into the exact electron density profile within the layered system must be simultaneously carried out in order to know the exact parameters influencing θ_0^P and R_{min}. Such an experimental method of course is the method of x-ray reflectivity. The specular x-ray reflectivity is traditionally used to characterise multi layer films [27].

With regards to thermal diffusion of gold colloids through a layer of polymer, the electron density normal to the layered system is measured after each annealing process. If the interface is not perfectly sharp and smooth then the reflected intensity will deviate from that predicted by the laws of Fresnel reflectivity. This deviation can then be analysed to obtain the electron density profile of the interface normal to the surface [28; 29]. The basic mathematical relationship which describes specular reflectivity is straight forward. When an interface is not perfectly sharp but has an average electron density profile given by $\rho_e(z)$, then the x-ray reflectivity can be approximated by:

$$R(Q)/R_F(Q) = \left| \frac{1}{\rho_\infty} \int_{-\infty}^{+\infty} e^{iQz} \rho_e(z) dz \right|^2 \qquad 3.15$$

Here R (Q) is the reflectivity, λ is the x-ray wavelength, ρ_∞ is the density depth within the thin film and θ is the angle of incidence. Typically one can then use this formula to compare parameterized models of the average electron density profile in the z – direction with the measured x-ray reflectivity and then vary the parameters until the theoretical profile matches the measurement. For films with multiple layers, x-ray reflectivity always shows oscillations. These oscillations can be used to infer layer thickness and other properties [27].

The information obtained from the x-ray reflectivity experiment (electron density profiles) and subsequent changes in the optical properties of the layered system (dielectric constant) obtain from the SPR spectroscopy are linearly dependent according equation (3.1) [30]

In the dynamic measurements of this work the optical (dielectric constant) and electron density variation changes during the thermal diffusion of colloids through a layer of polymer will be studied simultaneously using both spectroscopic methods mentioned above, device a model to help explain the experimental results obtained from the x-ray reflectivity method and finally use this model to verify the results obtain from the SPP plots.

From 3.9 it follows that in the angle of dispersive experiment the intensity maxima appears whenever $e^{-2ikT} = 1$ at an angle positions α_{im}. This condition can be expressed in the following way:

$$2T\sqrt{\sin^2\alpha_{im} - \sin^2\alpha_C} = m\lambda \qquad 3.16$$

This is similar to Bragg equation but modefied by the influence of refraction. The appearing thickness Fringes are called Kiessig fringes. A plot of $\sin^2\alpha_m$ as a function of the corresponding square of the m value is a straight line with y-intercept $\sin^2\alpha_C$ and gradient $\lambda^2/4T^2$, from whence the critical angle for the layer and the thickness can be determined. In the small angle limit, this equation can be used to determine the exact thickness of the layer and also corresponding modification can be made in the case of multiple layers.

In the theoretical account listed above, we have seen how one can use information obtained from x-ray measurements to calculate thickness of thin films using (39). With the simulation program XOP [13], data collected from measurements can be fitted into simulated models. In this way we can obtain additional parameters such as surface roughness of the thin film and the precise electron density distribution. With this additional information, the experimental trends obtained with the independent surface plasmons resonance response can be verified.

4 Experimental Methods

4.1 Sample characterisation techniques

A major part of this work deals with the characterisation of thin films either of varied electron density (alkenes, alcohols) or varied hydrogen concentrations (varied pH). Surface plasmons spectroscopy is a prominent optical method for the characterisation of dielectrics at surfaces and as a method of choice in this study permits the detection of such processes on metal substrates and is therefore described in some detail. On disadvantage of the simple surface plasmons resonance specstroscopy however is, that the sensitivity is not sufficient to characterise small molecular weight adsorbates or samples with low surface coverage. Secondly surface plasmons specstroscopy is only qualitatively able to give insight into the topology of a surface. It is in this light that a combination with x-ray reflectometry is inevitable if such information is desired.

4.1.1 Surface Plasmons Spectroscopy

In the following some experimental issues of surface plasmons specstroscopy are presented beginning with the different types of measurement modes. Starting from the basic angular mode, the time dependent is sketched with which adsorption kinetics, molecular switching behaviour or other time dependent surface processes can be studied. The second part is concerned with the experimental set-up of the normal surface plasmons resonance version and its various extensions

4.1.1.1 Measurement modes

4.1.1.1.1 Angular Dependent Measurements

As already mentioned, the basic measurement mode is represented by the resonance spectrum, i.e. normalised reflectivity versus angle of incidence, also referred to as a scan curve. The angle range of such a scan is important, as it has to cover the edge of total reflection, from which the refractive index of the medium can be determined, and the resonance minimum. The maximum range is limited by the geometry of our set-up which does not allow the motor to move beyond 90° as this may result to damages due to contacts with other parts of the device. However, samples used throughout the experiments had their resonance angle far before the 90° position of the goniometer. The light source used was a laser diode (Thorlab Inc., type CPS 192) with a wavelength of 780 nm and a band width of approximately 5nm. Prior to each scan, a polarizer is initially used to obtain the desired polarisation of the beam in our case the transverse magnetic component, p-polarised. The light, after being polarised is then directed onto the BK7 prism combined with a 60 nm Au film. The reflected beam of light is the reflected onto a photo diode detector (optoelectronics, model 220D) consisting of an active area of 10x20mm. The measurable angle in this configuration is then varied in steps of up to 0.01.

The amplification of measured signal is enhanced further with the help of amplifiers designed by the electronic department of the physics faculty. All the necessary equipment is controlled and read by the computer program Lab View. The obtain scan curves are then fitted into Lorentzian functions in order to obtain the thickness and or refractive index of the metal and dielectric layers. Measured results were later verified with the aid of the simulation program WINSPALL [56], which is based on the transfer matrix method. The parameters that have to be provided to the program are the measured reflectivity, the angle of incidence, the thickness and the dielectric constants of all layers as well as the laser wavelength and the geometry of the coupling prism. By iterative optimisation of the parameters the simulation can also be fitted to the scan curves to determine the optical constants of the involved layers or simply calculated from these curves. As already mentioned, the thickness and the refractive index of a layer cannot be determined separately by a single scan. However the refractive index of the surrounding medium can be extracted by fitting the edge of total reflection.

Considering surface plasmons spectroscopy as a standard method in biochemistry mostly the prism configuration is implemented in commercially available systems. This is due to several reasons: A flat surface offers a well defined geometry for chemical reactions that are to be investigated. Frequently, the specific binding or adsorption behaviour of the molecule is of interest. In the case of prism coupling the electric field of the surface plasmons and not the one of the incident light, interacts with the solvent. The exciting light impinges from the prism side and tunnels through the thin gold layer into the medium to be probed. The optical response of the probed medium e.g. a multiple layer can easily be calculated using Fresnells equations. There are however situations whereby the excitation of surface plasmons by grating can be favourable: By choosing an appropriate grating pitch, the resonance is observed only a few degrees of the normal incidence [31-46]. However one of the major disadvantages of the grating coupling technique is the analysis of the data. The rigorous simulation of the data is based on the complete characterisation of the grating as mentioned in (2.1.2.2.2). This procedure is extremely time consuming and requires much computer power. An easier, yet more accurate procedure is to convert the data measured by means of grating coupling into fictive prism coupling data, which can then be easily evaluated with the transfer matrix method. The excitation of surface plasmons always requires momentum matching of the incident light and the excitation and does not depend on the way (prism, grating) the momentum of the light is produced. Assuming that the coupling with the surface plasmons does not change the momentum of the light the only change in momentum due to a dielectric layer is the same for both experimental methods. The angle of incidence of a grating $\theta_{grating}$ can then be converted to those of an introduced fictive prism experiment θ_{prism}:

$$k_g + n_d.k_0.\sin\theta_{grating} = k_{x,SP} = n_p.k_0.\sin\theta_{prism} \qquad 4.1$$

k_g is the absolute value of the reciprocal grating vector, n_d and n_p the refractive indices of the dielectric and the prism respectively and k_0 is the absolute value of the wave vector in vacuum. In reality additional loss channels due to the finite thickness of the gold layer cannot be neglected. It has actually been showed [47] that despite these short comings, the conversion evaluation method still gives the right value within an error of less than 1% provided the grating amplitude is sufficiently small.

4.1.1.1.2 Time Dependent Measurements

Time dependent measurements or kinetic measurements represent the second type of measurement modus. Here the change in the dielectric or changes in the thickness are monitored as a function of time. Two different types of modi can be distinguished:
The first type measures the reflectivity at a fixed angle. In order to get reasonable data in such a modus it is important that the reflectivity around the fixed angle be a linear function of the incident angle. Therefore the fixed angle has to be chosen to be in the linear decreasing part of the resonance curve. Two assumptions are made: First a change in the dielectric constant as well as a change in the thickness of the dielectric layer or an additional layer causes a shift in the angle. Secondly it is implicitly assumed that the resonance curve its self does not changes its form but only shifted to different angles. If these assumptions are satisfied then the change

in reflectivity can be converted into refractive index or thickness changes and the initial and final values can be determined from the scan curve prior and after a kinetic measurement. The advantage of this type of measurement is the high time resolution. Since the incident angle is kept fixed, the resolution only depends on intergration time of the amplifier and typical values are 0.5 s to 5 s.

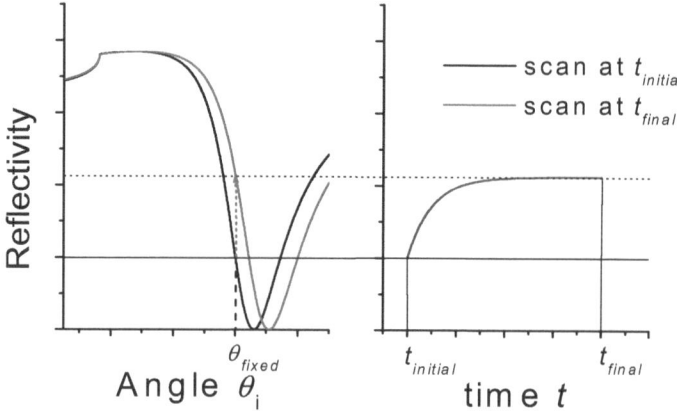

Figure 4.1: The left graph shows two scan curves measured before and after adsorption. In this case gold before and after adsorption of P3HT, at a fixed angel

In the second mode [22], the reflectivity is not measured at a fixed angle but rather at the angle of minimum of the resonance curve. An algorithm determines the reflectivity at an angle which is pre-set and supposed to be the minimum angel. Later two reflectivity values equally distant from the first value are taken. Through these three points a parabola is fitted. The angle at the minimum of the parabola serves as a starting angle at which the next data point is taken. In the case of very fast processes, a correction algorithm ensures that the goniometer moves further onto lower reflectivities until a parabola fit is feaseable. Thus, the method follows the actual minimum of the resonance curve and the fitting parabola additionally gives some information on the half-width of the resonance. Furthermore the procedure doesnot require the assumption of linearity as it is necessary in the first method. Another advantage is that any angle shift can be recorded whereas the former method is limited to shifts not larger than the half width of the resonace curve. In comparison with the fixed angle measurement the disadvantage of this method is the low time resolution. For each value three angle positions have to be adjusted and the amplifier always has to be read out. Depending on the hardware, this can take up to 6s to 9s.

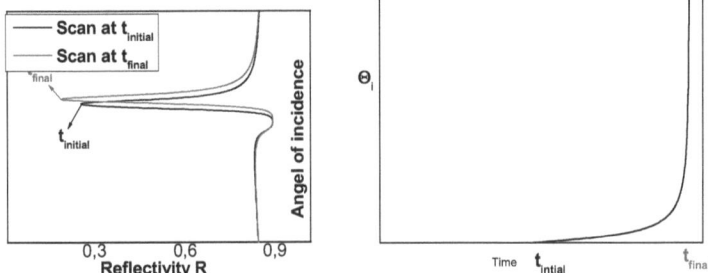

Figure 4.2: Right graph: Angular scan at $t_{initial}$ and t_{final}. Note the swapped axis. Left graph: Typical adsorption plot for a dielectric recorded at the resonance minimum.

4.1.1.2 Measurement Setup

All variations of the SPR set up discussed in this section are based on the fundamental set up scheme, which will be referred to as normal SPR set-up. Starting from this normal set up, different extensions will be introduced in the respective subsections.

4.1.1.2.1 Surface Plasmon Spectroscopy

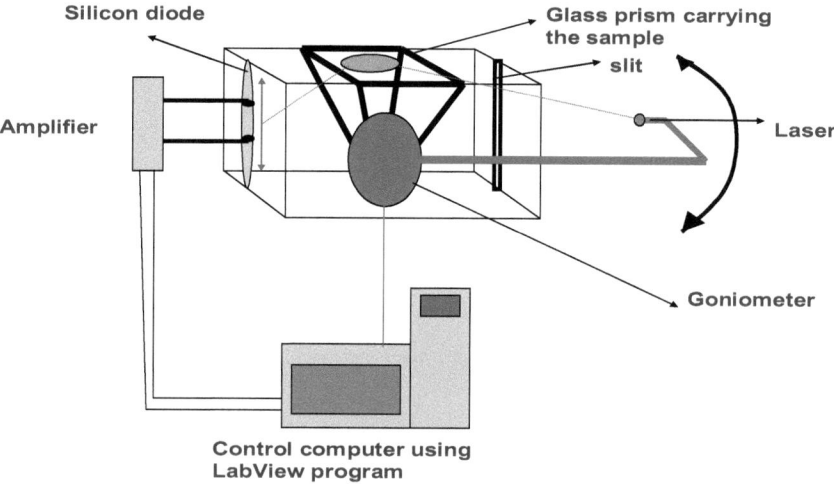

Figure 4.3: SPR spectrometer layout showing the various connections.

Figure 4.3 depicts the basic surface plasmons specstroscopy set-up. In general a laser with a wavelength of 780 nm (uniphased) is used. Prior to running any experiment, the beam is lead through a polarizer to ensure the right polarisation. The laser is attached to a goniometer which is programmed via Lab-View. The laser beam then passes through a slit and hits the sample (prism) which is mounted on rectangular aluminium box. Through Lab-View, the desired angle range can be set automatically. The goniometer is controlled by a step motor which provides a 1° /100 resolutions. After interacting with the sample, the out going beam is directed to a silicon diode with a relatively large surface area. A locked in amplifier finally reads the signal of the detector, which is proportional to the intensity of the reflected light, integrating over many oscillations. Typical oscillation times are 50-250 ms.

The glass prism used here is a BK7 glass prism with a refractive index of 1.55024. The metal carrying side of the prism can also be adapted to run flow experiments.

In case the metal responsible for the generation of surface plasmons isn't vaporised directly onto the surface of the prism, the outer side of the prism is index matched to the coupling prism and the gold can simply be evaporated on the same BK7 glass slices. This facilitates measurements particularly in the case were diverse measurements have to be carried out. The index matching fluid prevents an increased and undesirable reflection of the exciting light at the two glass/air interfaces between both glasses. In some of our experiments, refractive index matching fluid (n=1.51, Cargille) is therefore a compromise between optimum index matching properties and minimum volatility.

4.1.1.2.2 X-Ray Reflectometry

X-ray reflectometry is a useful tool to study thin films and multiple layered architecture from 2nm to about 5000nm. The typical x-ray wavelength used is of the order of an Angstrom ($Cu:K_\alpha = 1.54 Å$). Since the refractive index of x-rays is slightly less than unity, total internal reflection occurs at very small angles of incidence measured from the surface. At angles above the critical angle, x-rays start to penetrate the sample. At the interface between two layers, the radiation is partly reflected and partly transmitted into the layer below.

Having a mono or multiple films on a substrate, the total thickness is reflected in the interference pattern known as Kiessig fringes [27]. These Kiessig-fringes can be fitted with Fresnel's formulas [27] to reveal information about the layer thickness (frequency of oscillations), surface roughness (attenuation of the fringe amplitude), electron density of the layers (also influences the amplitude of the resonances) and the electron density of the surface layer (position of the critical angle).

Figure 4.4: Home laboratory x-ray device.

The intensity range measured in a reflectivity curve typically covers six or more orders of magnitude so a detector with a high dynamic range, usually a scintillation counter, is needed. In general a line source of x-rays is used such that at grazing incidence, a very large area (some cm²) of the sample is illuminated. Conventional x-rays reflectometers deploy the couple motion of the sample and the point detector. More elaborate reflectometer models use line detectors instead of point detectors and no moving parts are necessary any more hence the measurements gets considerably faster [48].

4.2 Sample Preparation Techniques

In the following the various techniques used to prepare and modify sample substrates are examined. All the surface plasmons resonance measurements were performed in the Kretschman-Reather coupling mode. No measurements were performed in the grating coupling mode. The techniques illustrated cover basic methods such as thermal everporation, self assembly of organic material and drop casting method.

4.2.1 Thermal Evaporation of Metal Layer

The metal gold needed for the surface plasmons experiments are thermally evaporated onto the glass prism. This process of thermal evaporation was carried out by the chemistry department. The films were prepared by directly depositing the gold (99.99%) of various thicknesses onto the glass substrate by an RF magnetron sputtering apparatus at a vacuum pressure of 3×10^{-6} Torr. The process was supervised by Herr Schulte. Varied thicknesses of gold were initially evaporated and plasmon curves of these bare surfaces were investigated. After considering the full width at half minima (FWHM) and intensity minima, a thickness of 50-60 nm was taken as optimal. The everporation rate was set at 0.1nm/s to 0.2nm/s. In order to enhance the adhesion of gold at the surface, a thin layer of chromium (2 nm 99.9%) was initially everporated on the surface.

4.2.2 Self Assembled Monolayers on Metal

Self-assembled monolayers (SAMs) are molecular assemblies that are formed spontaneously if an appropriate substrate is immersed into a solution of an active surfactant in a solvent [49; 50]. The popularity of these layers stems from the fact that well defined and closely packed assemblies can very easily be prepared at ambient laboratory conditions and that the strong coordination of the head group of the SAMs to the metal yields a layer that is sufficiently stable to desorption. Moreover the physical and chemical properties can be tailored by the choice of the end functional group for example $-COOH$, $-OH$, $-CH_3$ and by varying the length of the alkyl chain. Our sample of self organic monolayer was an organo-thiol, an organic acid attached to a thiol with the general formula HS-R-COOH. This was prepared in the University of Bochum. Due to the strong influence of x-rays on SAMs many experiments involving these types of molecules are usually carried in a vacuum. The absence of a vacuum in our home laboratory lead to a non-detection of these layers using x-rays however their chemical presence was detected using plasmons specstroscopy as we shall see later. From the thermodynamic point of view several parameters promote the assembly process on the surface. Chemisorption of the head-group of the surfactant leads to a strong attractive and exothermic reaction at the surface. Consequently, all the available binding sites on the surface are occupied. Additionally attractive van der Waals interaction between the alkyl group can stabilise the molecule. In this work only a single thiol was used although there many other examples like alcohols on gold or carboxylic acids on aluminium oxide.

4.2.3 Drop Casting

The drop casting technique is a rather not so efficient technique particularly when films of a particular thickness are needed. Another shortcoming of this technique is that the uniformity of the film cannot be guaranteed. The technique involves inclining the surface over which the layer is to be deposited and with the help of a pipette a few drops of the film in question are dropped on the inclined surface and left to run down. For experiments where film uniformity doesn't play a vital role or influence the results, this method is particularly good. In most of the experiments the gold layer was deposited directly on the prism hence spin coating wasn't possible by virtue of the relative thickness and weight of the sample. It's because of these short comings that we resorted to drop casting.

5 Static Measurements

5.1 Detection of alkanes, alkynes, alcohols and organic acids:

Alkanes, alkynes, alcohol and water all have different refractive indices and since surface plasmons are sensitive to refractive index changes on the surface of the metal generating them, it should be possible therefore to detect this variation using our home made device. In investigating this variation, the sample in question was brought on the surface of the active gold metal surface and the step angle was then changed to 1°/2s to avoid total evaporation of the samples (in the case of methanol, ethanol and hexane) before the end of the experiment.

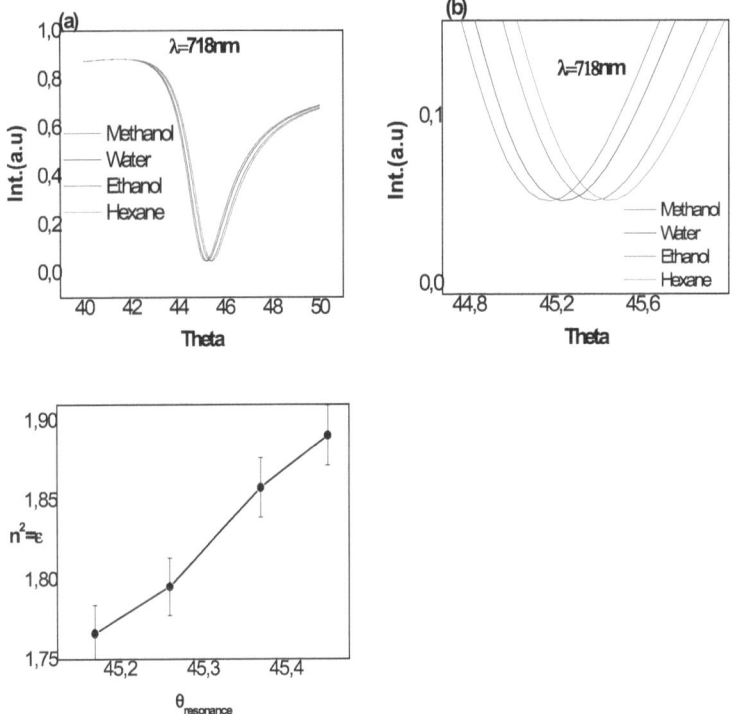

Figure 5.1: Resonance angle variations for methanol, water, ethanol and hexane.

In this experiment basically the functionality of the SPR sensor was tested. It is absolutely vital that the gold sensing surface is clean because the system does respond to the presence of contaminants even in very small quantities.

Before every experiment, the surface is cleaned using ethanol, which is dispensed on top of the surface using a dropper and then blow dried with a fan. Figure 5.1c illustrates an almost linear relationship of the dielectric constants as a function of the resonant angle.
This is what we normally expect because theoretically, the resonant angle is related to the dielectric of the surface as follows:

$$\sin\theta_{res.} = \frac{1}{n_p}\left[\frac{\varepsilon_m \varepsilon_s}{\varepsilon_m + \varepsilon_s}\right]^{\frac{1}{2}} \qquad 5.1$$

At resonance the change in the angle position of the minima is very small hence we can write $\sin\theta_{res.} \approx \theta_{res}$. This then modifies equation (1) to:

$$n_p^2 \theta_{res.}^2 = \frac{\varepsilon_m \varepsilon_s}{\varepsilon_m + \varepsilon_s} \qquad 5.2$$

Differentiating (5.2) with respect to ε_s yields:

$$\frac{d\theta_{res.}}{d\varepsilon_s} = \frac{(\varepsilon_m + \varepsilon_s)\varepsilon_m - \varepsilon_m \varepsilon_s}{2n_p^2 (\varepsilon_m + \varepsilon_s)^2} \qquad 5.3$$

This expression can still be written in the reciprocal form i.e.

$$\frac{d\varepsilon_s}{d\theta_{res.}} = \frac{2n_p^2 (\varepsilon_m + \varepsilon_s)^2}{(\varepsilon_m + \varepsilon_s)\varepsilon_m - \varepsilon_m \varepsilon_s} = \frac{2n_p^2 (\varepsilon_m + \varepsilon_s)^2}{\varepsilon_m^2}$$

$$\frac{d\varepsilon_s}{d\theta_{res.}} = 2n_p^2 \left[1 + \frac{\varepsilon_s}{\varepsilon_m} + \left(\frac{\varepsilon_s}{\varepsilon_m}\right)^2\right]. \qquad 5.4$$

This is a linear relation if $\left(\frac{\varepsilon_s}{\varepsilon_m}\right)^2 \approx 0$ which is the case in our experiment since $\varepsilon_m = -22.4$ and $\varepsilon_s \approx 1.4$.

The table below summarizes the results of the resonance angles and the corresponding indices of refraction:

Specimen	Resonance angle	Refractive index
Methanol	45,17°	1,329
Water	45,26°	1,340
Ethanol	45,37°	1,363
Hexane	45,45°	1,375

Table 5.1: SPR angle and refractive indices for methanol water ethanol and hexane

Despite the promising results of the integrated SPR sensor, there are certainly numerous avenues that can be explored to improve the accuracy, repeatability, and detection limit, such that the performance is comparable to commercial SPR systems. Some suggestions are listed below:

– the noise sources in the system has to be characterize more thoroughly and each source must be targeted independently to improve noise performance of the system.

– Development of a more robust post-processing technique to locate the SPR dip position and filter out any noise if possible.

– A big improvement could be made if the design of the device was modified slightly to move the input incidence angle out of the operating range of the sensor.

– Scale down the device. The current dimensions of the system are very conservative and were chosen to facilitate fabrication and testing. However, since the operating concept has been tested and verified, the device can be scaled down by a factor of 2 times, leading to a substantially miniaturized device.

5.2 Effect of wavelength on the resonance angle

Most commercial instruments use curves of the reflected intensity vs. angle at a single wavelength (670 nm for the Ibis II from Holland Biomaterials, 760 nm for Biacore instrument, 840 nm for Texas Instrument's Spreeta system) and either assume the optical index of bio-molecules to be known and extrapolable ($n_{protein} \approx 1.41$ in most cases) or calibrate their instrument using an independent technique applied to a limited range of species (e.g. Biacore using radioactive labelling [54]). The assumption on the optical index which is required for analyzing single wavelength SPR measurements is a major drawback when unknown samples are used [54].

De Bruijn et al [51] have previously analyzed the limitations of SPR measurements and described iterative algorithms for identifying these properties in a more general context (requiring multiple measurements with various thicknesses of layers made of the unknown dielectric material).

In order to verify the functionality of our device to varied wavelength, surface plasmons plots were recorded for the same sample using varied lasers.

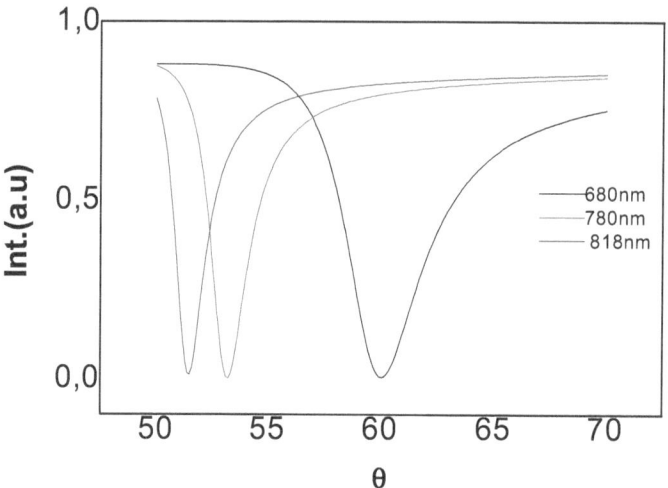

Figure 5.2: Resonance angle as a function of wavelength. Sample system consisted of Hexane on a 60 nm thick gold layer.

The sample in question was hexane and laser sources of the following wavelength were used: 680 nm, 780 nm and 818 nm. The table below summarises the variation of the resonance angle as a function of the wavelength used simulated using the computer program WINSPALL. We see consistency with our experimental results.

Wavelength/nm	Resonance Angle/°	
	Experimemt	Simulation
680	59.0	60.10
780	52.0	53.10
818	50.5	51.50

Table 5.2: Variation of the resonance angle as a function of the wavelength.

At resonance, the relationship between the resonance angle and the dielectric constant of the added film is given according to (5.1) above.

To understand the trend in the experimental results we must first of all understand the variation of the real and imaginary components of the dielectric constant to wavelength. The figure below shows this trend:

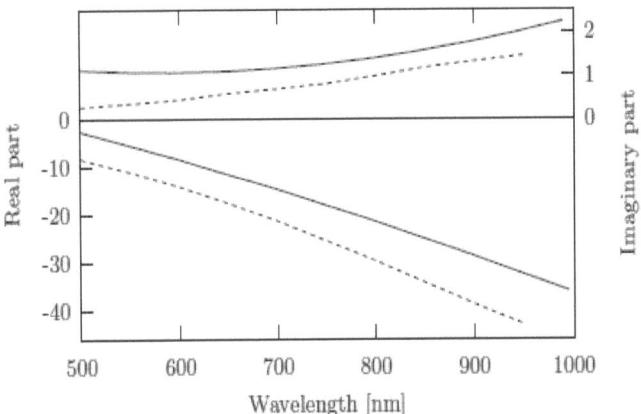

Figure 5.3: Variation of the real and complex component of the dielectric function with wavelength. [53]

From the figure one can see that for a given specimen, the real part of its dielectric function decreases while its imaginary part increases with increasing wavelength. This means with respect to our experiment that by increasing the wavelength, the denominator of equation (5.1) reduces steadily. Although the numerator also reduces, it can be shown that $x*y > x+y$ implying that there is a net decrease in 5.1. This inverse relationship of the dielectric constant with the wavelength therefore explains the shift to lower resonance angle as we increase the wavelength.

5.3 Simultaneous detection of thickness and dielectric constant of self assembled monolayer

Self-assembly is the basic principle which produces structural organization. It is a reversible processes in which pre-existing parts of a pre-existing system form structures of patterns or mechanisms. Self-assembly can be classified as either static or dynamic. Static self-assembly is when the ordered state does not dissipate energy and is in equilibrium while dynamic self-assembly requires dissipation of energy.

Self-assembly is scientifically interesting and technologically important at least for four reasons. The first one is that it is critically important in life. The cell contains a great range of complex structures such as lipid membranes, folded proteins, molecular machines, structured nucleic acids, and many others that form by self-assembly. Secondly, self-assembly provides straight forward path to a range of materials with regular structures: molecular crystals, liquid crystals are examples. Third, self-assembly also occurs commonly in systems of components larger than molecules. Fourth, self-assembly seems to offer one of the most general strategies now available for generating nanostructures. Therefore, self-assembly is important in a range of fields: chemistry, physics, biology, materials science, nano-science, and manufacturing [11].

Self-assembly is a natural way to make things. In order to finding efficient ways to make micro-machines and faster electronic devices, scientists are working on self-assembly to put things together with something like "natures' ease"[10].

Self-assembly is an important part of nanotechnology because using tools to manipulate objects in molecular scale is very difficult and not cost-effective.

In light of this numerous advantages, our aim is to use our device and see if it's possible to detect the presence of SAMs on gold substrate using the surface plasmons resonance and x-ray reflectivity spectroscopy methods simultaneously.

The sample in question, an organothiol, with general molecular formula HS-R-X (R: alkane group, X: carboxylic group) was produced by the chemistry department of the university of Bochum. The BK7 glass plate over which the gold layer has been evaporated was dip into a solution containing the HS-R-X. This was then allowed in a still stand condition for two days During this time period the SAMs gradually forms on the surface of the gold layer.

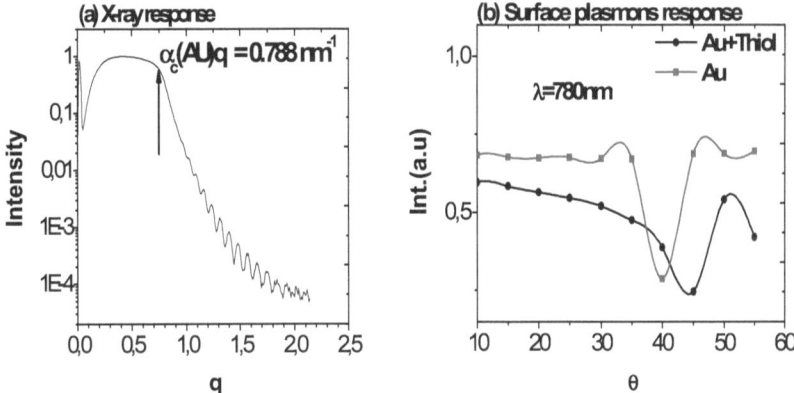

Figure 5.4: (a) X-ray response; (b) Surface plasmons response of a gold film carrying an organothiol.

The surface plasmons resonance response shows a shift of the resonance from 41.0° (for the unmodulated gold surface) to 44.4° (organothiol on gold surface). From this shift in resonance angle the value of the dielectric constant of the added layer was found to be 1.17.

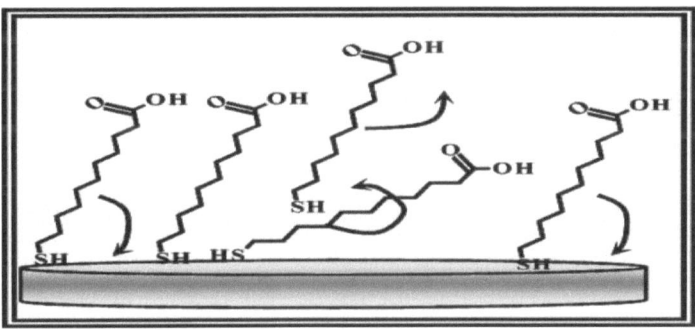

Figure 5.5: Self assembled organothiol on gold.

This value is different from the experimental value of the dielectric constant of this SAM at room temperature which is 9.08 [55].

Since the exact length of the organic thiol is known to be 1-3 nm, one has to calculate the total thickness of the gold layer after submersion into the thiol solution. From the x-ray results, we can see only one critical angle which corresponds to that of the gold layer. This is a signal that there is the absence of an additional layer over the gold film. Using the relationship $T = 2\pi / \Delta q_z$, one comes out with a thickness of 47.9 nm. The thickness of the gold film determined from x-ray reflectivity was 48 nm. This implies that the SAMs were destroyed or blown off by the beam of the x-rays. However their presences on the surface of the gold film lead to an alteration of the electron density on the surface of the gold film. The surface plasmons resonance device was therefore capable to detect this electron density variation; however the x-ray device could not detect these fallen strings of the SAMs. The destruction of SAMs by x-rays has necessitated the carrying out of such experiments under vacumm conditions.

6 Dynamic Measurement

The dynamic diffusion of colloids through a membrane plays a vital in many processes. Osmosis in plants is a process whereby a stronger solution draws a relatively weaker concentrated solution through a semi-permeable membrane. This process is responsible for the absorption and transportation of ions from the soil to the leaves of plants.

The application of drug-smugglers which in the case of cancer therapy is the adhesion of taxon on gold colloid and its subsequent transportation to the diseased cells requires that the precise sizes and the diffusion process of these colloids be understood. Subsequent excitation of these colloids with near infra red radiation leads to the generation of surface Plasmon which then kill the diseased cells.

These diverse applications of colloids motivated our study of their diffusion across a polymer layer deploying both the process of surface Plasmon resonance and small angle X-ray reflectivity.

6.1 Thermal Dynamic Diffusion of Gold Colloids through P3HT

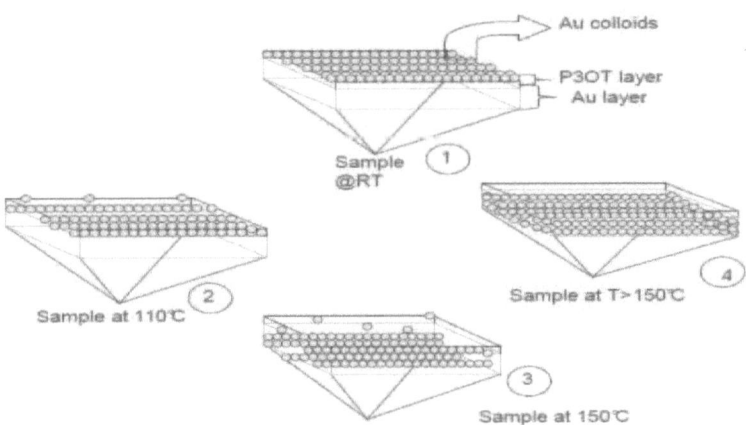

Figure 6.1: Annealing process of colloids through P3HT.

The diagram above shows the various steps associated with the thermal diffusion of gold colloids through a polymer layer. A 60 nm thick gold film was thermally evaporated onto the surface of a glass prism. Later a layer of P3HT was drop casted onto the gold film and the total thickness (gold film plus polymer) was determined using x-ray reflectivity measurements. Through subtraction, the thickness of the polymer film was found to be 45nm. Gold colloids of diameter 20 nm were then drop casted onto the polymer matrix and the system was then heated stepwise from room temperature to 150°C, 170°C, 200°C and later 220°C for five minutes. From room temperature when the colloids are on the surface of the polymer, the sample is heated to a temperature above the glass temperature of the polymer. This leads to the diffusion of the colloids through the polymer layer assuming all other factors which can cause diffusion of the colloids into the polymer layer are neglected. Subsequent annealing leads to more and more colloids diffusing till when there are no more colloids at the surface of the polymer.

After each phase of the annealing process, simultaneous x-ray small angle reflectivity and surface Plasmon resonance spectrum are recorded. Regarding the SPR spectra, the incident angle was varied insteps of 0.1° from 0° to 50° and the out going intensity was measured by a silicon diode (Optoelectronics, model 220D). The x-ray spectra was measured using a scintillation counter in steps of 0.01° from 0° to 3°.

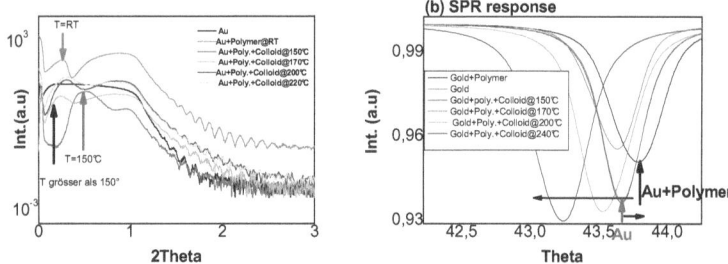

Figure 6.2: (a) X-ray response, (b) SPR before and after annealing the sample.

First we discuss the results of SPR measurements.

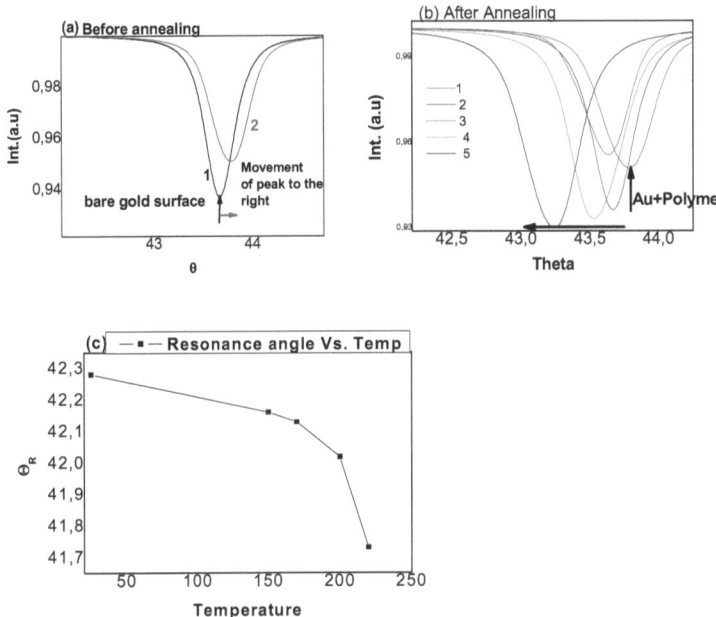

Figure 6.3: Angular shift of the SPR curves (a) before annealing (1) bare gold surface, (2) gold/P3HT; (b) after covering Au colloids onto P3HT layer and annealing (1) bare gold surface (2) Au/P3HT/colloid at 150°C, (3) Au/P3HT/colloid at 170°C , (4) Au/P3HT/colloid at 200°C, (5) Au/P3HT/colloid at 220°C; (c) Resonance angle as function of temperature.

The SPR response shows a shift of the resonance angle prior to annealing first to the right. With reference to eq. (5.1), this implies that after covering the P3HT thin film onto the gold surface the dielectric constant $\varepsilon_2 = \varepsilon_{air} = 1$ is changed to $\varepsilon_{P3HT} > 1$. This larger dielectric constant is detected as a change in the average electron density of the gold film leading to an increase in θ_R according to (5.1), i.e. a shift of the resonance peak minimum to the right (higher angle).

Upon heating the sample from 150°C onwards, the electron density profile within the polymer layer is varied due to diffusion of the gold colloids through the polymer layer. This diffusion results in a shift of θ_R to smaller angles as shown in figure 5b.

The diffusion of the colloids through the polymer layer brought about a variation in the electron density profile within the film. This varying density profile has been determined independently by x-ray reflectivity measurements as shown in figure 22a.

From fig.22a, at room temperature (see pink curve) the x-ray reflectivity measurement reveals thickness fringes and two critical angles; one at $2\theta = 0.25°$ corresponding to P3HT and the other at $2\theta = 1.1°$ corresponding to gold film.

At higher temperatures, e.g. T=150°C (red curve), the scans show a shift of the critical angle of the polymer film to the right, implying that the colloids have diffuse within the polymer layer causing the initial critical angle to move to higher values. Further annealing from 170°C to 220°C moves the peak towards smaller angle positions again (see blue curve, for example). This means that gold colloids diffuse gradually into the polymer/gold interface leading to a gradual reduction of the electron density within the polymer film compared to the electron density at T=150°C. The trend in the variation of the vertical concentration of gold colloids appearing after each step of annealing can be determined via x-ray reflectivity simulation. This can be achieved by sub-dividing the polymer layer into a set of sub-layers (in this case nine sub-layers). By varying the chemical constitution, density and surface roughness of each sub-layer, the electron density variation is simulated. This procedure is repeated for each sub-layer until the theoretical profile matches the measurement. Before getting into that we must first of all carry out theoretical calculations in order to know the limits of the new dielectric constant brought about by the mixture of P3HT and gold colloids during the annealing procedure.

6.1.1 Theoretical consideration

To fully understand this trend in response, the following model was suggested:

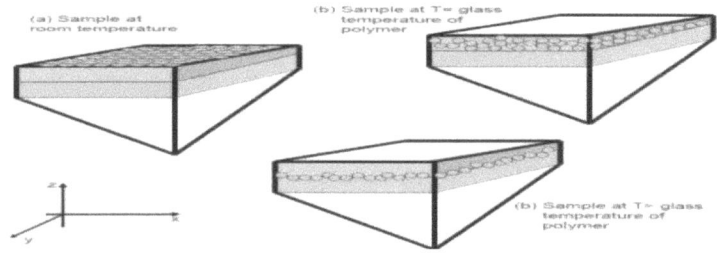

Figure 6.4: Theoretical model from experiment.

Neglecting the effects of diffusion brought about due to gravity and assuming that the diffusion of the colloids is caused principally by the effects of thermal perturbation, the model suggests that after heating the sample, the colloids start diffusing into the polymer and form a gold/polymer complex. By subsequently increasing the temperature the diffusion process increases until when all the colloids subsides at the gold/polymer interface. This means that after each annealing procedure, the effective dielectric constant of the polymer/colloid complex changes due to the diffusion of colloids. We therefore would like to know the range of the possible values of the new dielectric constant during this procedure.

We will assume surface plasmon propagation in the positive x direction on the metal dielectric boundary lying in the x-y plane, with the fields tailing off in the into the positive (dielectric) and negative (metal) direction described by the following TM waves (indices 1 and 2 denote dielectric and metal regions respectively):

$$E_j = (E_x, 0, E_{z_j}) \exp(i(k_x x + k_{z_j} z - \omega t))$$
$$H_j = (0, H_y, 0) \exp(i(k_x x + k_{z_j} z - \omega t))$$
$$j = 1, 2$$
6.1

From Maxwell's equations and continuity at the boundary, the following relationships are derived:

$$\frac{\varepsilon_1}{k_{z_1}} = \frac{\varepsilon_2}{k_{z_2}}$$
$$k_x^2 + k_{z_i}^2 = \varepsilon_i k_0^2$$
$$E_x = \frac{k_{z_i}}{\omega \varepsilon_i} H_y; \quad E_{z_i} = -\frac{k_x}{\omega \varepsilon_i} H_y$$
$$i = 1, 2$$
6.2

Here $k_0 = \frac{\omega}{c}$ is the free space wave vector of the incident excitation vector. From equation (6.2) the surface plasmons dispersion relation can be derived:

$$k_x^2 = k_0^2 \frac{\varepsilon_1 \varepsilon_2}{\varepsilon_1 + \varepsilon_2} \quad (a)$$
$$k_{z_i}^2 = k_0^2 \frac{\varepsilon_i^2}{\varepsilon_1 + \varepsilon_2} \quad (b)$$
6.3

The imaginary part of k_x is responsible for the lossy propagation of the surface plasmon along the interface. This results from the non-zero imaginary component of the metal permittivity. Apart from choosing a metal with a high plasmonic resonance $\frac{\varepsilon_2'}{\varepsilon_2''} \gg 1$, there seems to be no other method to reduce the metallic losses so as to increase the surface plasmons propagation length in the metal dielectric configuration.

As we heat the sample, the gold colloids start diffusing into the polymer as evident by the x-ray reflectivity results. This intends leads to a modification of the chemical architecture of the polymer layer.

Replacing the passive dielectric medium in region 1 by a dielectric medium with gain will enable us to compensate for the losses of the metal as evident by the calculation below. We will start by assigning to region (1) a gain medium with complex permittivity $\varepsilon_1' + i\varepsilon_1''$, the complex component being contributed by the diffused gold colloids and investigate the condition for a bound wave to propagate at the interface. At this point our only assumption is that ε_2' is negative and its absolute value is much larger than the other three permittivity components. The equations governing the surface plasmons propagation for this material configuration are:

$$k_x^2 = k_0^2 \frac{(\varepsilon_1' + i\varepsilon_1'')(\varepsilon_2' + i\varepsilon_2'')}{(\varepsilon_1' + \varepsilon_2') + i(\varepsilon_1'' + \varepsilon_2'')} \quad (a)$$

$$k_{z_1}^2 = k_0^2 \frac{(\varepsilon_1' + i\varepsilon_1'')^2}{(\varepsilon_1' + \varepsilon_2') + i(\varepsilon_1'' + \varepsilon_2'')} \quad (b)$$

6.4

First we find the limits of ε_1'' for a bound solution (i.e. $\text{Im}(k_{z_1}) > 0$). From Eq. (6.4b) we have:

$$k_{z_1} \approx k_0 \sqrt{\frac{|\varepsilon_1' + \varepsilon_2'|}{(\varepsilon_1' + \varepsilon_2')^2 + (\varepsilon_1'' + \varepsilon_2'')^2}} \left(-\varepsilon_1'' + i\varepsilon_1' \right)\left(1 - i\frac{(\varepsilon_1'' + \varepsilon_2'')}{2(\varepsilon_1' + \varepsilon_2')} \right)$$

6.5

From Eq. (6.5) and the condition $\text{Im}(k_{z_1}) >$ we have:

$$(\varepsilon_1'')^2 + \varepsilon_2'' \varepsilon_1'' + 2\varepsilon_1'(\varepsilon_1' + \varepsilon_2') < 0$$

6.6

This is satisfied if:

$$\frac{-\varepsilon_2^{'} - \sqrt{(\varepsilon_2^{"})^2 - 8\varepsilon_1^{'}(\varepsilon_1^{'} + \varepsilon_2^{'})}}{2} < \varepsilon_1^{"} < \frac{-\varepsilon_2^{'} + \sqrt{(\varepsilon_2^{"})^2 - 8\varepsilon_1^{'}(\varepsilon_1^{'} + \varepsilon_2^{'})}}{2} \qquad 6.7$$

The condition on the size of $\varepsilon_2^{'}$ guarantees that these bounds are real and have opposite signs. This in turn places a limit on the allowable amount of absorption or gain in the dielectric for bound waves to exist. Substituting values of $\varepsilon_2^{"}(Au)$ and $\varepsilon_1^{'}(Polymer)$ in Eq. (6.7) yield the following limits for the complex component of the new dielectric: $-6,23 < \varepsilon_1^{"} < 7,98$.

The longitudinal propagation characteristics of the surface plasmons are given by k_x in Eq. (6.4a), which simplifies to:

$$k_x^2 = \frac{k_o^2}{(\varepsilon_1^{'} + \varepsilon_2^{'})^2 + (\varepsilon_1^{"} + \varepsilon_2^{"})^2}\left[\varepsilon_1^{'}((\varepsilon_2^{"})^2 + \frac{|\varepsilon_1|^2}{\varepsilon_1^{'}}\varepsilon_2^{'} + (\varepsilon_2^{'})^2) + i\varepsilon_1^{"}((\varepsilon_1^{"})^2 + \frac{|\varepsilon_2|^2}{\varepsilon_2^{"}}\varepsilon_1^{"} + (\varepsilon_1^{'})^2)\right] \qquad 6.8$$

where $|\varepsilon|^2 = (\varepsilon^{'})^2 + (\varepsilon^{"})^2$. Assuming $\varepsilon_1^{'}(polymer), \varepsilon_2^{'}(Au)$ and $\varepsilon_2^{"}(Au)$ are fixed by the choice of metal and polymer respectively, we solve for $\varepsilon_1^{"}$ to find the roots of the imaginary part of Eq. (6.8), which has to be zero for lossless propagation of surface plasmons polariton, yielding:

$$\varepsilon_1^{"} = \frac{|\varepsilon_2|^2}{2\varepsilon_2^{"}}(-1 \pm \sqrt{1 - \frac{4(\varepsilon_1^{'}\varepsilon_2^{"})^2}{|\varepsilon_2|^4}}) \simeq \begin{cases} \frac{|\varepsilon_2|^2}{2\varepsilon_2^{"}} + \frac{(\varepsilon_1^{'})^2 \varepsilon_2^{"}}{|\varepsilon_2|^2} \quad (a) \\ -\frac{(\varepsilon_1^{'})^2 \varepsilon_2^{"}}{|\varepsilon_2|^2} \quad (b) \end{cases} \qquad 6.9$$

Here the approximation is valid assuming a metal with a high plasmonic resonance. We also observe that the sign of $\varepsilon_1^{"}$ is opposite to the sign of $\varepsilon_2^{"}$. This implies that there is gain in region 1. The magnitude of the solution in Eq. (6.9a) is too large and outside the bounds given in Eq. (6.7), so we will only consider Eq. (6.9b). Substituting values of $\varepsilon_1^{'}(polymer)$ and $\varepsilon_2^{"}(Au)$ in Eq. (6.9b), we obtain another limit for $\varepsilon_1^{"}$ namely $\varepsilon_1^{"} > 3,8*10^{-3}$.

In this algebraic consideration, we have used the condition for the existence of a bound wave on the new configuration to find a limit for the new complex component of the dielectric constant starting with Eq. (6.6).

We will still use this equation and form it in terms of ε_1' to find the limits of its value for this new configuration.

Re-written in terms of ε_1', Eq. (6.6) now takes the following takes the following form:

$$2\left(\varepsilon_1'\right)^2 + 2\varepsilon_1'\varepsilon_2' + \varepsilon_1''\left(\varepsilon_1'' + \varepsilon_2''\right) < 0 \qquad 6.10$$

Solving for ε_1' we obtain the following:

$$\varepsilon_1' < \frac{-\varepsilon_2' + \sqrt{(\varepsilon_2')^2 - 2\varepsilon_1''(\varepsilon_1'' + \varepsilon_2'')}}{2} \qquad 6.11$$

By substituting we obtain the following limit for the new real component of the dielectric constant: $\varepsilon_1' < 22,4$.

Here we observe again that the sign of the new ε_1' is opposite to that of gold and slightly less in magnitude.

In the aforementioned analysis, we have been able to obtain limits for the new permittivity of our new configuration. It must be mentioned however that the orientation of the colloids within the polymer also plays a vital role to the magnitude of the new dielectric constant. For a polymer layer of thickness 45 nm there exist varied possibilities by which colloids of diameter 20 nm can be arranged within as shown in the illustration below.

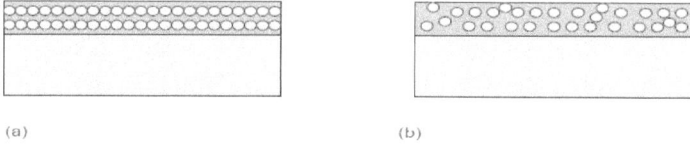

(a)   (b)

Figure 6.5: Colloid distribution within polymer layer showing ordered (a) and random distribution (b) of colloids.

Based on the above distribution patterns, the effective dielectric constant of the new configuration can be further approximated as a set of paralleled arranged layers with as shown above with configuration (a) the least probable. However we are going to use this configuration as an onset to estimate the effective dielectric constant of the configuration. A set of parallel capacitors has an effective capacitance given by the following expression:

$$\frac{1}{\varepsilon_{effective}} = \frac{1}{\varepsilon_1' + i\varepsilon_1''} = \sum_{i=1}^{n}\left[\frac{1}{\left[\varepsilon_{Colloid}' + i\varepsilon_{Colloid}''\right]}\right]_i + \sum_{i=1}^{n}\left[\frac{1}{\varepsilon_{Polymer}'}\right]_i \qquad 6.12$$

Here the first part of the right hand side is the contribution from the colloids while the second part is the contribution from the polymer.

The possible numbers of terms is deduced by considering how many possible gold colloids of a given diameter can fit on the polymer layer if vertically arranged. In the case of colloids with diameter 20nm and a polymer layer of 45nm, one can deduce that only two colloid layers are possible. Therefore the effective dielectric simplifies to an expression with contributions from 2 colloids layers and 3 polymer sub-layers in between and this can be expressed as follows:

$$\frac{1}{\varepsilon_1' + i\varepsilon_1''} = \frac{\left(2\varepsilon_{poly}' + 3\varepsilon_{Au}'\right) + 3i\varepsilon_{Au}''}{\left(\varepsilon_{Au}' + \varepsilon_{poly}'\right) + i\varepsilon_{Au}''} \qquad 6.13$$

After rationalising the numerator of the right hand side, we obtain the following expression for the real and complex component of the new configuration

$$\frac{1}{\varepsilon_1' + i\varepsilon_1''} = \frac{\left(\varepsilon_{Au}' + \varepsilon_{poly}'\right)\left(2\varepsilon_{poly}' + 3\varepsilon_{Au}'\right) + 3\left(\varepsilon_{Au}''\right)^2}{\left(2\varepsilon_{poly}' + 3\varepsilon_{Au}'\right)^2 - \left(3\varepsilon_{Au}''\right)^2} - i\frac{\varepsilon_{Au}''\varepsilon_{poly}'}{\left(2\varepsilon_{poly}' + 3\varepsilon_{Au}'\right)^2 - \left(3\varepsilon_{Au}''\right)^2} \qquad 6.14$$

The algebraic analysis above has given us a clue into the range and the signs of the new dielectric which results from a mixture between gold colloids and P3HT assuming the colloids align themselves parallel within the polymer layer.

Substituting the known values for the dielectric constant for gold and P3HT, we obtain the following range for the values of the new dielectric constant assuming parallel alignment

$\varepsilon'_{new} \leq 5.47; \varepsilon''_{new} \leq 5.3*10^{-3}$. This will serve as a guide when carrying out the simulation of the x-ray results to obtain the electron density variation within the polymer which will further lead us into calculating the various dielectric constants at every given temperature.

6.1.1.1 Simulation of x-ray measurements using the simulation program XOP

From fig.22a, at room temperature (see pink curve) the x-ray reflectivity measurement reveals thickness fringes and two critical angles; one at $2\theta = 0.25°$ corresponding to P3HT and the other at $2\theta = 1.1°$ corresponding to gold film.

At higher temperatures, e.g. T=150°C (red curve), the scans show a shift of the critical angle of the polymer film to the right, implying that the colloids have diffuse within the polymer layer causing the initial critical angle to change. Further annealing from 170°C to 220°C moves the peak towards smaller angle positions again (see blue curve, for example). This means that gold colloids diffuse gradually into the polymer/gold interface leading to a gradual reduction of the electron density within the polymer film compared to the electron density at T=150°C. The trend in the variation of the vertical concentration of gold colloids appearing after each step of annealing can be determined via x-ray reflectivity simulation. This can be achieved by sub-dividing the polymer layer into a set of sub-layers (in this case nine sub-layers). By varying the chemical constitution, density and surface roughness of each sub-layer, the electron density variation is simulated. This procedure is repeated for each sub-layer until the theoretical profile matches the measurement. In simulating the experimental results obtain from x-ray spectroscopy, the program XOP version 2.1[57] was used. First the bare gold surface is simulated by varying the surface roughness and thickness parameters until the theoretical profile fits measurement. The same procedure is used to simulate the bi-layer system P3HT on gold.

However with the inclusion of gold colloids on the surface of the polymer and subsequent annealing, the electron density within the polymer layer varies from the surface towards the gold/prism interface. To model this variation in the electron density, the polymer layer is divided into nine sub-layers of equivalent thickness but varied electron density .After each annealing procedure the concentration of gold within the gold colloids-polymer complex is varied by changing the sub-layer density, surface roughness and chemical constitution until the corresponding simulation result fits the experimental scan. The electron density variation within the entire polymer is then later plotted and fitted with a Gaussian function.

The glass substrate in all cases was modelled using silicon dioxide. The disparity in the alignment of the Kiessig fringes between experiment and simulation arises from the extreme high surface roughness of the polymer/colloid complex.

Table 6.1: Prism/Gold

Density	Concentration of Au	Thickness T/nm	Roughness (σ)
19.75	100%Au	60	18
Substrate:	SiO2		4

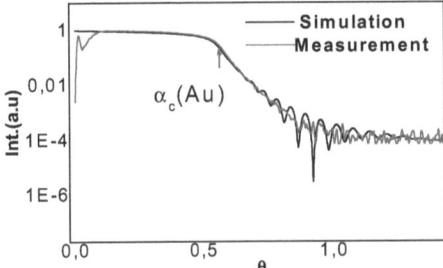

Figure 6.6: Simulation and experimental measurement of the gold layer.

Agreement with experiment is achieved considering a 60nm thick gold layer on top of a glass substrate modulated by silicon dioxide.

Table 6.2: Prism/Gold/ P3HT at room temperature:

Density ($\rho[g/cm^3]$)	Electron density ratio	Thickness T/nm	Roughness (σ)
1.4	100%P3HT	45	80
19.75	100%Au	60	19

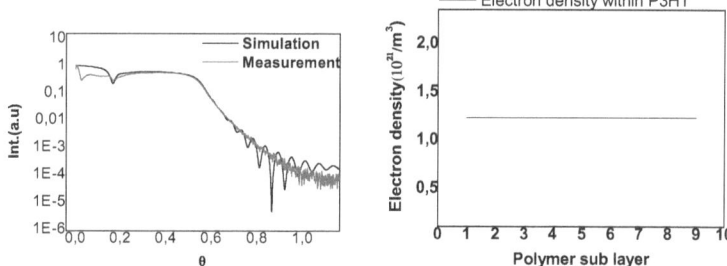

Figure 6.7: Simulation and experimental measurement of the layer structure prism/gold/P3HT at room temperature.

Table 6.3: Prism/Gold/P3HT/colloid at 150°C

Density ($\rho[g/cm^3]$)	Electron density ratio	Thickness T/nm	Roughness (σ) in nm
1.4	100%P3HT	5	10
15	1 atom of Au/ P3HT molecule 50%Au;50%P3HT	5	10
16	2 atoms of Au/P3HT molecule 66%Au;33%P3HT	5	10
19.75	20 atoms of Au/P3HT molecule 95%Au ;5%Polymer	20	10
19	2000 atoms of Au/P3HT molecule 99.9%Au;0.1%P3HT	5	10
18	10 atoms of Au/ P3HT molecule 90%Au;10%P3HT	5	10
19,75	100% Au	60	65
Substrate:	SiO2		

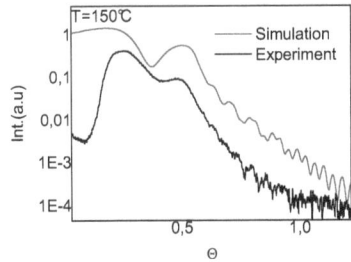

Figure 6.8: Simulation and experimental measurement of the layer structure prism/gold/P3HT/Colloids at T=150°C (b).

Figure 6.8 shows the results for the diffused gold into the polymer after annealing at 150°C for 5 minutes. The angle positions of both layers are well reproduced. Discrepancies in intensity for angles $\theta < 0.2°$ can be explained by the primary beam effect which was not considered in the simulation.

In order to calculate the electron density variation within the polymer layer from the results obtain from the x-ray reflectivity simulation, we will assume a mono atomic layer of gold colloids on the surface of the polymer layer with density $\rho = 10*10^3 kg/m^3$. We will also assume a thickness of 2 mm and the surface area of the film to be $1 cm^2$. This then yields a total of $6*10^{21}$ gold colloid particles on the surface of the polymer layer. This number of particles will diffuse within the polymer in the course of the annealing procedure. From the x-ray simulation results, each sub-potion of the polymer layer has an electron density which consists partly of the contribution of the gold colloids and partly from the polymer. To calculate the contribution by gold, we use the various electron density ratio of gold and then calculate the number of gold particles found in each polymer sub-layer.

Table 6.4: Gold particle distribution within polymer at T=150°C

Polymer sub-layer	% Au	Number of Particles
1	0	0
2	50	$0.438*10^{21}$
3	66	$0.578*10^{21}$
4	95	$0.832*10^{21}$
5	95	$0.832*10^{21}$
6	95	$0.832*10^{21}$
7	95	$0.832*10^{21}$
8	99	$0.867*10^{21}$
9	90	$0.788*10^{21}$

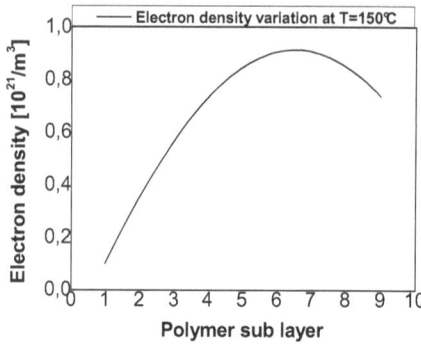

Figure 6.9: Electron density variation within polymer at T=150°C

Table 6.4: Gold/P3HT/colloid at 170°C

Density ($\rho[g/cm^3]$)	Electron density ratio	Thickness T/nm	Roughness (σ)
1.4	100%P3HT	5	Total roughness for all three sub-layers 60
1.4	100%P3HT	5	
1.4	100%P3HT	5	
1.5	25%Au;75%P3HT	5	15
1.7	33.3%Au;66.6%P3HT	5	10
1.8	50%Au;50%P3HT	15	10
1.7	33.3%Au;66.6%P3HT	5	4
19.75	100%Au	60	11
Substrate :	SiO2		

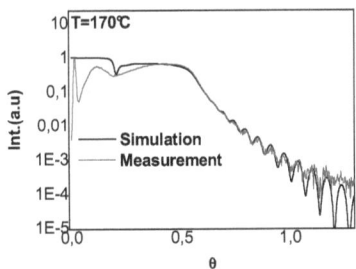

Figure 6.10: Simulation and experimental measurement of the layer structure prism/gold/P3HT/Colloids at T=170°C.

By annealing the sample again to 170°C for another five minutes, the distribution of gold colloids within the polymer layer changes with respect to the distribution at T=150°C due to further diffusion. This has as a direct consequence the movement of the critical angle of the polymer layer to a relatively lower angle value (0.12°) with respect to the value at 150°C (0.26°). The visibility of Kiessig oscillations proves the homogeneity of the layer.

We will again use the ratio of the electron density to calculate the number of particle for each sub polymer layer at this temperature in order to obtain the new electron density variation.

Table 6.5: Gold particle distribution within polymer at T=150°C

Polymer sub-layer	% Au	Number of Particles
1	0	0
2	0	0
3	0	0
4	25	$0.622*10^{21}$
5	33	$0.821*10^{21}$
6	50	$1.245*10^{21}$
7	50	$1.245*10^{21}$
8	50	$1.245*10^{21}$
9	33	$0.822*10^{21}$

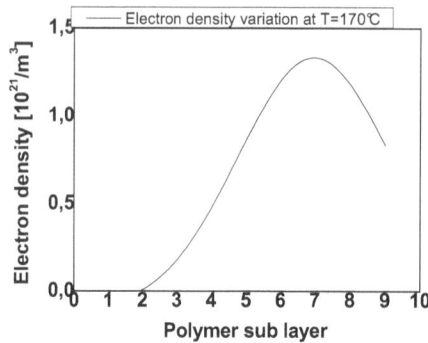

Figure 6.11: Electron density variation within polymer layer at T=170°C

Table 6.6: Prism/Gold/Polymer/Colloid at 200°C

Density ($\rho[g/cm^3]$)	Electron density ratio	Thickness T/nm	Roughness (σ)
1,4	100%P3HT	5	Total Polymer roughness 70
1.4	100%P3HT	5	----------------
1.4	100%P3HT	5	----------------
1.4	100%P3HT	5	----------------
1,6	75%Au;25%P3HT	5	15
3.2	66.6%Au;33.3%P3HT	15	70
1.7	33.3%Au;66.6%P3HT	5	4
19,75	100%Au	60	12
Substrate:	SiO2		

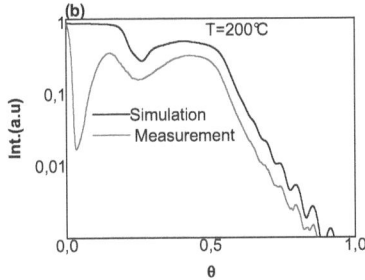

Figure 6.12: Simulation and experimental measurement of the layer structure prism/gold/P3HT/Colloids at T=200°C

The sample is further annealed to 200°C for five minutes. Peak position of the polymer layer remains invariant. Due to homogeneity of the layer system, we see again the presence of Kiessig oscillations. The new distribution of colloids within the polymer layer is calculated using the ratio of the electron density of gold from the simulation results.

Table 6.7: Gold particle distribution within polymer at T=170°C.

Polymer sub-layer	% Au	Number of Particles
1	0	0
2	0	0
3	0	0
4	0	0
5	75	$1.47*10^{21}$
6	66	$1.29*10^{21}$
7	66	$1.29*10^{21}$
8	66	$1.29*10^{21}$
9	33	$0.647*10^{21}$

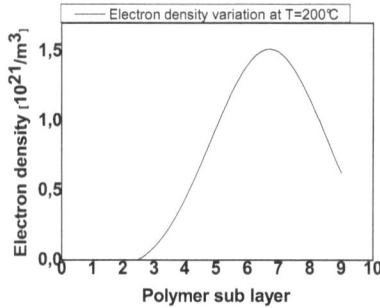

Figure 6.13: Electron density variation at T=200°C

Table 6.8: Prism/Gold/Polymer/Colloids at 220°C

Density ($\rho[g/cm^3]$)	Electron density ratio	Thickness **T/nm**	Roughness (σ)
1.4	100%P3HT	20	Total polymer roughness 70
1,4	100%P3HT	5	_____
1.4	100%P3HT	5	_____
1.4	100%P3HT	5	_____
3.5	50%Au;50%P3HT	5	70
4	66.6%Au;33.3%P3HT	5	4
19.75	100%Au	60	12
Substrate:	SiO2		

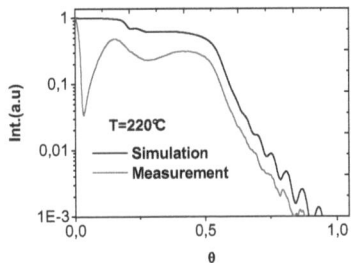

Figure 6.14: Simulation and experimental measurement of the layer structureprism/gold/P3HT/Colloids at T=220°C (b).

By raising the temperature again to 220°C for another seven minutes, the polymer peak position doesn't show any visible changes with respect to the situation at 200°C. The homogeneity again is noticeable by the presence of oscillations.

The new distribution of colloids is again calculated by the same method as in the previous situations.

Table 6.9: Gold particle distribution within polymer at T=200°C

Polymer sub layer	%Au	Number of particles
1	0	0
2	0	0
3	0	0
4	0	0
5	0	0
6	0	0
7	0	0
8	50	$2.586*10^{21}$
9	66	$3.41*10^{21}$

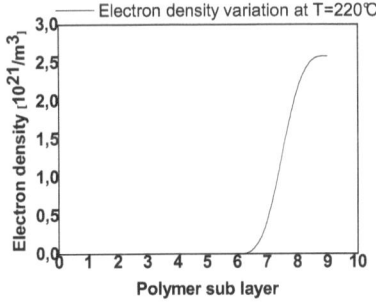

Figure6.15: Electron density variation within polymer at T=220°C

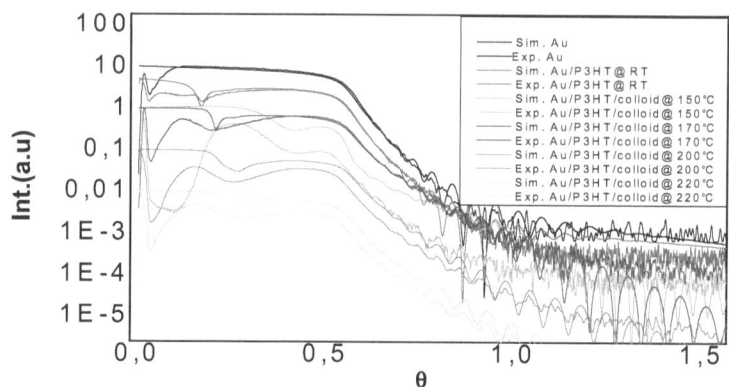

Figure 6.16: Simulation and experimental results of x-ray reflectivity.

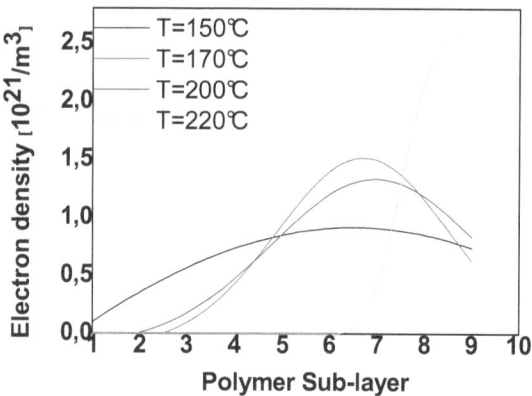

Figure 6.17: Electron density variation within the polymer layer as a function of temperature.

The X-ray results for both experiment and simulation are shown above. From the simulated x-ray results, the electron density variation within the polymer layer was calculated and plotted as a function of each sub-layer of the polymer layer. After annealing the sample to 150°C for five minutes, the subsequent diffusion of gold colloids into the polymer layer is seen on electron density variation profile within the polymer layer at this temperature. This reveals a relatively higher electron density profile. Neglecting all other factors which might bring about diffusion into the polymer implies at this temperature the initial influx of colloids into the polymer begins. On the x-ray results, we notice a movement of the critical angle of the polymer peak from 0.12° (room temperature) to 0.26° (150°C). By further annealing the sample to 170°C for another five minutes, we observe on the x-ray results a movement of the critical angle position of the polymer towards lower angles position. This implies a variation in the electron density within the polymer film. As we further anneal up to 220°C, the concentration of the colloids within the polymer layer drifts towards the polymer gold interface. We see this on the electron density spatial drift towards the base of the polymer gold interface.

6.1.1.2 Simulation of surface plasmons curve using X-ray model

In simulating the SPR responses using the electron densities obtained from the x-ray model above, one needs to calculate the dielectric constant from the calculated electron density for each given sub-layer. The real and complex components of the dielectric constant and the electron density are related to each other according to the following relationships [58]:

$$\varepsilon'(\omega) = \left(1 - \frac{\omega_p^2}{\omega^2}\right) = \left(1 - \frac{\rho e^2}{4\pi^2 f^2 \varepsilon_0 m}\right)$$

$$\varepsilon''(\omega) \approx \frac{\omega_p^2}{\omega^3 \tau} = \frac{\rho e^2}{8\pi^3 f^3 \varepsilon_0 m \tau}$$

6.15

Here e is the electronic charge, m is the mass of an electron, ω_p represents the plasma frequency, ω is the frequency of the laser source, τ is the collision time and ρ represents the electron density.

Again we will commence with the prism/gold surface, prism/gold/polymer and after the inclusion of colloids on the surface of the polymer and subsequent heating, the electron density variation within the polymer layer obtained from the x-ray reflectivity simulation will be used to calculate the dielectric constant for each sub-layer. With the calculated values of the dielectric constant, the simulation program WINSPALL design by scientists at the Max-Planck institute of material science in Mainz was employed to carry out the simulation [56]. In the simulation program, at resonance almost all of the incident beam couples to the surface plasmons implying a fall in the intensity of the reflected beam to almost zero. Due to our experimental set up however the intensity of the SPR signal does not go to zero as modelled by WINSPALL. A combination of experimental and simulation results in one diagram therefore is improbable due to high variations in intensities. Like in any SPR plot, the resonance angle and the full width at half minima (FWHM) are very decisive. Hence at the end of the modelling, a comparison of these parameters for both experiment and simulation will be used to ascertain the consistency between experiment and simulation.

Table 6.7: Prism/Gold

layer thickness T /nm	ε'	ε''
60	-22.4	1.4

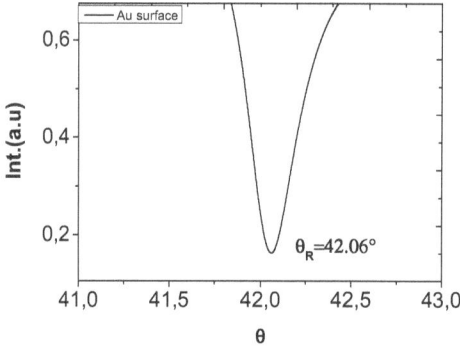

Figure 6.18: SPR simulation of gold surface on prism.

The bare gold surface on glass was model with a layer of gold of the same thickness on a triangular prism. No considerations for the surface roughness were made in the simulation. The values for ε' and ε'' correspond to the dielectric constants of gold at $\lambda = 780nm$.

Table 10: Gold/P3HT at room temperature

layer thickness T /10^{-10}	ε'	ε''
P3HT 450	1.06	0
Au 600	-22.4	1.4

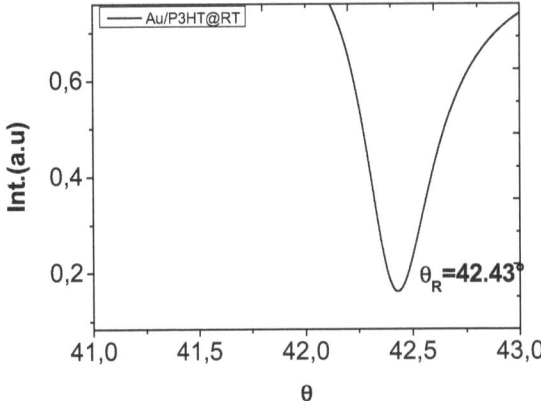

Figure 6.19: SPR simulation of the layered structure prism/Au/P3HT at room temperature.

The bi-layer system gold/P3HT on BK7 prism substrate was modelled in the same way as the non-modulated gold surface. The complex component of the dielectric constant of P3HT is zero hence no absorption. The additional P3HT film has caused a movement of θ_R from 42.06° (bare gold surface) to 42.43° (gold/P3HT) implying an increase in the electron density on the surface of gold.

Table 6.11: Gold/P3HT/colloids at 150°C

layer thickness T /nm	Electron density ratio [1/m³]	ε'	ε''
5	100%P3HT	1.06	0
5	50%Au;50%P3HT	1	0.00209
5	66%Au;33%P3HT	1	0.00586
20	99.9%Au;0.1%P3HT	1	0.066
5	66%Au;33%P3HT	1	0.00523
5	99.9%Au;0.1%P3HT	1	0.02466
60	Au	-22,4	1,4

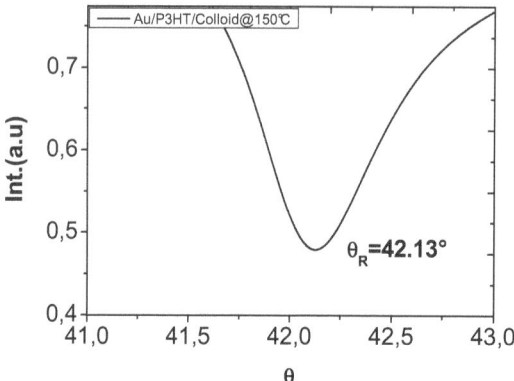

Figure 6.20: SPR simulation of the layered structure prism/Au/P3HT/colloids at T=150°C.

From T=150°C onwards, the respective values of the dielectric constant for each sub-layer are calculated using equations 6.15. A leftward movement of the resonance angle also sets in at this temperature onwards.

Table 6.12: Gold/P3HT/Colloids at 170°C

layer thickness T /nm	Electron density ratio	ε'	ε''
5	100%P3HT	1.06	0
5	100%P3HT	1.06	0
5	P3HT	1.06	0
5	25%Au;75%P3HT	1	0.000003
5	33.33%Au;66.66%P3HT	1	0.0000034
15	50%Au;50%P3HT	0.98	0.0000037
5	33.33%Au;66.66%P3HT	1	0.0000034

| 60 | Au | -22.4 | 1.4 |

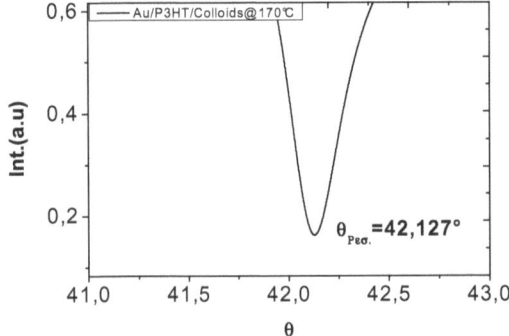

Figure 6.21: SPR simulation of the layered structure prism/Au/P3HT/colloids at T=170°C.

At 170°C the spatial as well as amplitude of the electron density variation within the polymer layer set in. The resonance angle has been shifted to the left. Here the spatial re-location of the electron density plays a vital role as we shall see later.

Table 6.13: Gold/Polymer/colloids at 200°C

layer thickness T /nm	Electron density ratio	ε'	ε''
5	100%P3HT	1.06	0
5	100%P3HT	1.06	0
5	100%P3HT	1.06	0
5	100%P3HT	1.06	0
5	75%Au;25%P3HT	1	0.0000037
15	66.66%Au;33.33%P3HT	0.92	0.000007
5	33.33%Au;66.66%P3HT	1	0.0000034
60	Au	-22.4	1.4

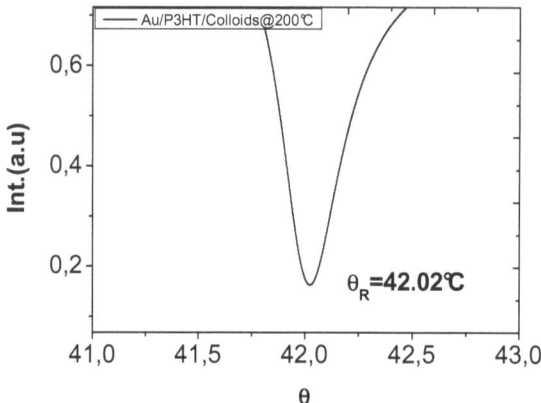

Figure 6.22: SPR simulation of the layered structure prism/Au/P3HT/colloids at T=200°C.

Annealing to 200°C for five minutes causes another shift of the resonance angle to the left. Again the spatial localisation of the electron density distribution within the polymer layer is accountable for this trend.

Table 6.14: Gold/P3HT/Colloids at 220°C

layer thickness T /nm	Electron density ratio	ε'	ε''
5	100%P3HT	1.06	0
5	100%P3HT	1.06	0
5	100%P3HT	1.06	0
5	100%P3HT	1.06	0
5	100%P3HT	1.06	0
5	100%P3HT	1.06	0
5	100%P3HT	1.06	0
5	50%Au;50%P3HT	0.70	0.0000072
5	66.66%Au;33.33%P3HT	0.71	0.0000088
60	Au	-22.4	1.4

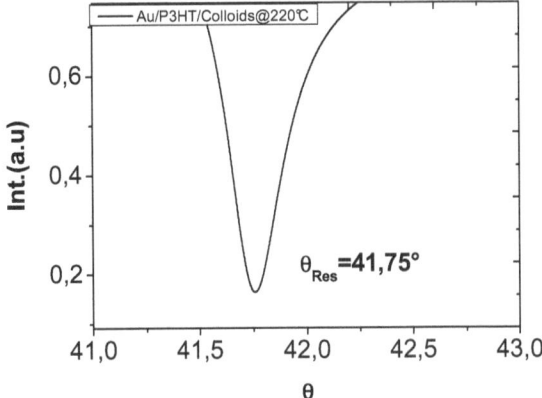

Figure 6.23: SPR simulation of the layered structure prism/Au/P3HT/colloids at T=220°C.

Further annealing to 220°C for another five minutes has further led to colloids finally settling at the interface polymer/gold layer.

As previously mentioned, due to intensity variations between experimental and simulation results, both results will be shown independently. The resonance angle however will be depicted on one graph as a function of temperature for both the experimental and simulation results as shown below.

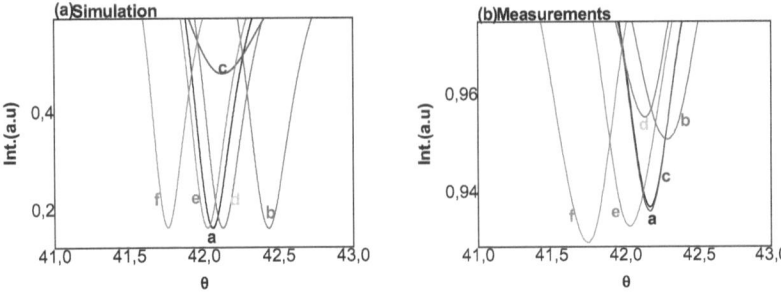

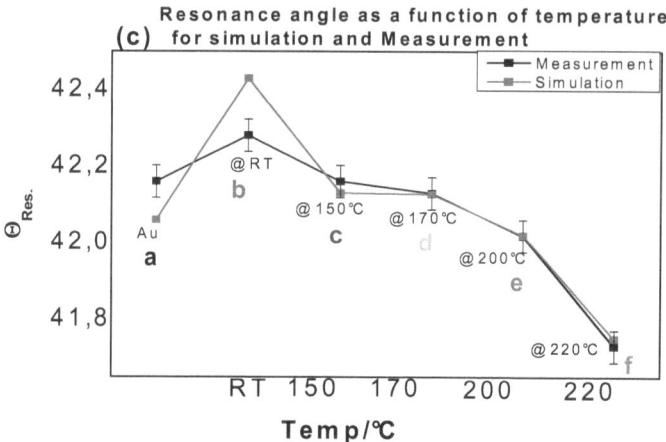

Figure 6.24: SPR responses (a) Simulation (b) Measurement (c) Resonance angle as a function of temperature for both experiment and simulation. For the graphs of both simulation and measurement the following is valid: (a) bare gold surface, (b) gold/P3HT, (c) gold/P3HT/colloid at 150°C, (d) gold/P3HT/colloid at 170°C, (e) gold/P3HT/colloid at 200°C, (f) gold/P3HT/colloid at 220°C.

6.1.1.3 Discussion

From the x-ray simulation results it is clear that the electron density within the polymer layer varies continuously after each annealing process due to the continuous diffusion of gold colloids through the polymer layer. This distribution has been approximated by a Gaussian function and shows an almost uniform electron density distribution at a temperature of 150°C within the entire polymer. At this temperature we also observe that the minimum of the SPR curve moves towards the position of that of the bare gold surface supporting the hypothesis that the polymer layer is entirely filled with gold. The amplitude as well as the spatial localisation of the electron density falls as the temperature is increased due to diffusion of the colloids towards the polymer/gold interface. From this trend it is evident that the colloids gradually diffuse through the polymer and this process of diffusion continues gradually towards the polymer/gold interface with increasing temperature.

In order to verify if the electron density distribution within the polymer layer is responsible for the trend in the SPR response (movement of the resonance peak to the left after each annealing sequence), an independent simulation with a layer of P3HT was carried out. To mimic the diffusing colloids, a thin layer of gold is placed within the polymer film and its thickness is systematically increased as it diffuses within the polymer layer.

A	B	C	D	E	F	G	H	I
P3HT (1nm)	P3HT (6nm)	P3HT (11nm)	P3HT (16nm)	P3HT (21nm)	P3HT (26nm)	P3HT (31nm)	P3HT (36nm)	P3HT (41nm)
Au (1nm)	Au (6 nm)	Au (11nm)	Au (16 nm)	Au (21nm)	Au (26nm)	Au (31nm)	Au (36nm)	Au (41nm)
P3HT (44nm)	P3HT (39nm)	P3HT (34nm)	P3HT (29nm)	P3HT (24nm)	P3HT (19nm)	P3HT (14nm)	P3HT (9nm)	P3HT (4nm)
Au (60nm)	Au (60nm)	Au (60nm)	Au (60nm)	Au (60nm)	Au (60nm)	Au (60nm)	Au (60nm)	Au (60nm)

Table 6.15: Simulation parameters for a layer of P3HT and a gold film.

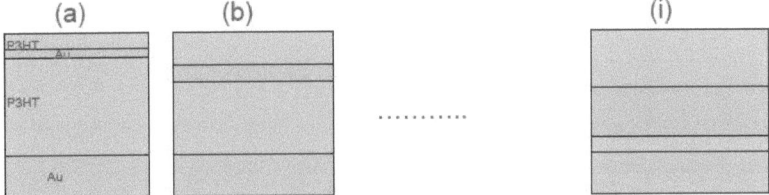

Figure 6.25: Simulation model (not drawn to scale) showing a thin gold film which gradually increases as it diffuses through the polymer layer. The thin gold film simulates the electron density

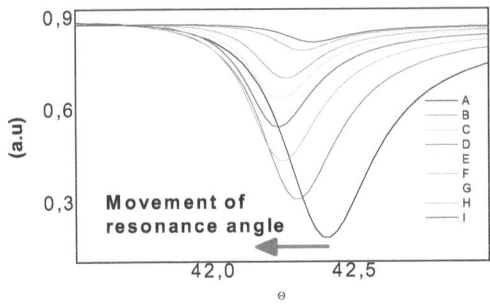

Figure 6.26: SPR response of the model parameters of table 6 showing a movement of the resonance angle towards the left as the gold film diffuses.

The simulation result clearly shows that the spatial localisation of the electron density within the polymer layer as well as the variation in its amplitude are the responsible factors for the movement of the resonance curve to the left after each annealing procedure.

To support this claim from our experimental findings, we must first of all superimpose the variation in the electron density over the evanescence wave function responsible for the SPR response. The evanescence wave decays exponentially into the polymer [59]. The effective penetration depth d is given by:

$$E_{ever.}(z) \sim e^{-z/d}$$

$$d = (k_{tz})^{-1} = \frac{\lambda}{4\pi\sqrt{n_1^2 \sin^2 \alpha - n_2^2}} \qquad 6.16$$

Here n_1 and n_2 denote the refractive indices of the two media, α represents the angle of incidence as measured from the normal to the surface and λ is the wavelength of the laser source. These surface-bound decaying waves are also the basis for many experimental techniques that characterise surface changes.

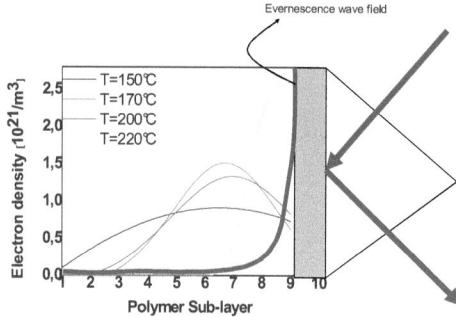

Figure 6.27: Superposition of evanescence wave field (red) over electron density distribution within polymer layer at T=150°C , 170°C, 200°C and 220°C.

One can observe from the above sketch that, although the magnitude of the electron density distribution function is extremely high at the initial annealing temperature (T=150°C), and relatively low in magnitude at higher temperatures (T=170°C-220°C), the exponential decaying nature of the evanescence wave (indicated by the red line) into the polymer and the spatial translation of the electron density towards higher evanescence wave field with increasing temperature guarantees that at higher temperatures the evanescence wave ''sees'' or couples with a relatively higher electron density on the polymer than it does at lower temperatures. Since $E_{ever.}(z) \sim e^{-z/d}$, it is clear that the coupling strength is highest at z=0 (immediate surface of the gold) and at the z=22.5nm (middle of polymer layer), the intensity has fallen to almost zero.

Therefore the high electron density at T=150°C is almost not detected since it is spatially located far away from the point where the evanescence wave field is maximum.

Theoretically the dielectric function for an inhomogeneous electron system is related to the electron density as in equation (5). Applying this relationship to the fact that at higher temperatures the evanescence wave couples with a relatively higher electron density implies a decrease in the local dielectric function is detected.

A decrease in the local dielectric function is seen on the surface plasmons spectroscopy response as coupling at lower angles. This explains why systematically increasing the temperature of the sample leads to a movement of the resonance curves towards the left as shown by the curves of the SPR spectroscopy.

A direct consequence in this peak movement resulting from the electron density variation within the polymer layer can be used in cell imaging and thermal therapy in the field of medicine. If properly miniaturised, such a device can detect chemical variations within a cell resulting from changes in the concentration of certain ions. The concentration of carbon dioxide in the atmosphere can also be monitored by such peak movements since the peak shift is a direct implication that the ionic concentration within the medium has increased. The food industry can also profit from this finding.

6.2 Non Diffusion of Colloids on P3HT

In this experiment, colloids of diameter 100 nm were brought onto the surface of the polymer matrix by the method of drop casting. Here the colloids were chosen such that they could not diffuse and mix with the polymer layer as in the prior case. The annealing time was again five minutes pro annealing step and after each annealing procedure, scans for both the SPR responds as well x-ray were recorded.

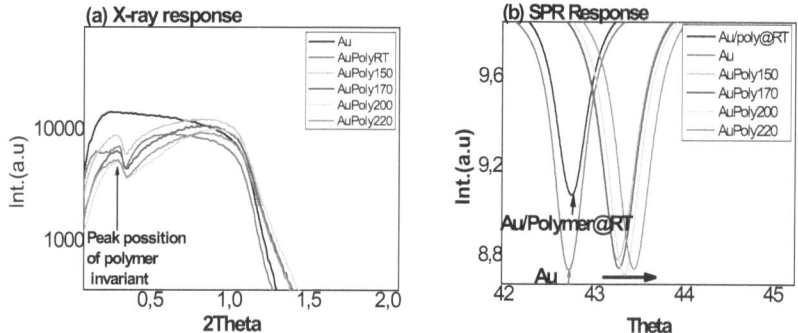

Figure 6.28: (a) X-ray response and (b) SPR response of 100 nm colloids on 45 nm film P3HT on the surface of gold.

The positional invariance of the peak position of the critical angle of the polymer in the x-ray experiment simply means that through out the annealing process the there wasn't any diffusion process of the colloids. This can be understood because the relatively lesser thickness of the polymer layer with respect to the colloid diameter (100 nm) cannot allow the colloids to diffuse through it.
The plasmon respond on the other hand shows that at higher annealing temperatures, the resonance coupling process is moved to higher angle.

6.2.1 Theoretical Consideration

Due to the fact that there wasn't any diffusion of the colloids through the polymer as evident by the x-rays results, we now turn our attention to surface plasmons in multiple layers of alternating conducting and dielectric thin films to explain the trend in the surface plasmons responds. Here we will assume that the gold colloids form a uniform layer on the surface of the polymer and then study the new modes resulting from the configuration. In such a system each single interface can sustain bound surface plasmons polaritons. When the separation between adjacent interfaces is comparable to or smaller than the decay length of the interface mode, interaction between surface plasmons polaritons give rise to coupled modes. In order to elucidate the general properties of coupled modes, we will focus on our specific three-layer geometry depicted on the figure below: a thin dielectric core (I) (in this case P3HT) sandwiched between two metallic claddings (II and III) i.e. a metal/insulator/metal hetero structure.

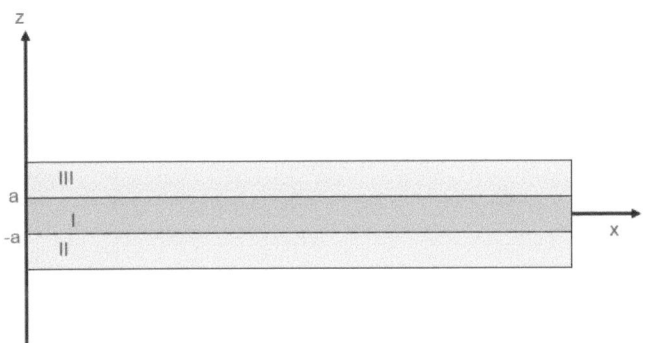

Fig. 6.29: Geometry of a three layer system consisting of a thin layer of P3HT (I) sandwiched between two layers of gold (II and III).

Since we are here only interested in the lowest order bound modes, we will start with a general description of transverse magnetic modes (TM) that are non-oscillatory in the z-direction normal to the interface.

For TM modes the systems of governing equations are:

$$E_x = -i\frac{1}{\omega \varepsilon_o \varepsilon}\frac{\partial H_y}{\partial z}$$
$$E_z = -\frac{\beta}{\omega \varepsilon_o \varepsilon} H_y$$

6.17

and the wave equation for TM modes is:

$$\frac{\partial^2 H_y}{\partial z^2} + \left(k_0^2 \varepsilon - \beta^2\right) H_y = 0$$

6.18

For z > a, the field components are:

$$H_y = A e^{i\beta x} e^{-k_3 z}$$
$$E_x = iA \frac{1}{\omega \varepsilon_0 \varepsilon_3} k_3 e^{i\beta x} e^{-k_3 z} \qquad 6.19$$
$$E_z = -A \frac{\beta}{\omega \varepsilon_0 \varepsilon_3} k_3 e^{i\beta x} e^{-k_3 z}$$

while for z < -a we get:

$$H_y = B e^{i\beta x} e^{k_2 z}$$
$$E_x = -iB \frac{1}{\omega \varepsilon_0 \varepsilon_2} k_3 e^{i\beta x} e^{k_2 z} \qquad 6.20$$
$$E_z = -B \frac{\beta}{\omega \varepsilon_0 \varepsilon_2} k_3 e^{i\beta x} e^{k_2 z}$$

Thus we demand that the fields decay exponentially in the claddings (II) and (III). Note that for simplicity we denote the component of the wave vector perpendicular to the interfaces simply as $k_i \equiv k_{z,i}$.

In the core region $-a < z < a$, the modes localised at the bottom and top interface couple, yielding

$$H_y = C e^{i\beta x} e^{k_1 z} + D e^{i\beta x} e^{-k_1 z}$$
$$E_x = -iC \frac{1}{\omega \varepsilon_0 \varepsilon_1} k_1 e^{i\beta x} e^{k_1 z} + iD \frac{1}{\omega \varepsilon_0 \varepsilon_1} k_1 e^{i\beta x} e^{-k_1 z} \qquad 6.21$$
$$E_z = C \frac{\beta}{\omega \varepsilon_0 \varepsilon_1} k_3 e^{i\beta x} e^{k_1 z} + C \frac{\beta}{\omega \varepsilon_0 \varepsilon_1} k_3 e^{i\beta x} e^{-k_1 z}$$

The requirement of continuity of H_y and E_x leads to:

$$A e^{-k_3 a} = C e^{k_1 a} + D e^{-k_1 a}$$
$$\frac{A}{\varepsilon_3} k_3 e^{-k_3 a} = -\frac{C}{\varepsilon_1} k_1 e^{k_1 a} + \frac{D}{\varepsilon_1} k_1 e^{-k_1 a} \qquad 6.22$$

at z = a, and

$$Be^{-k_2a} = Ce^{-k_1a} + De^{k_1a}$$

$$-\frac{B}{\varepsilon_2}k_2e^{-k_2a} = -\frac{C}{\varepsilon_1}k_1e^{-k_1a} + \frac{D}{\varepsilon_1}k_1e^{k_1a} \qquad 6.23$$

at $z = -a$, a linear system of four coupled equations.

H_y further has to fulfil the wave equation in the three distinct regions, via:

$$k_i^2 = \beta^2 - k_0^2\varepsilon_i \qquad 6.24$$

for $i = 1, 2, 3$. Solving this system of linear equations results in an implicit expression for the dispersion relation linking β and ω via

$$e^{-4k_1a} = \frac{\frac{k_1}{\varepsilon_1} + \frac{k_2}{\varepsilon_2}\Big/\frac{k_1}{\varepsilon_1} + \frac{k_3}{\varepsilon_3}}{\frac{k_1}{\varepsilon_1} - \frac{k_2}{\varepsilon_2}\Big/\frac{k_1}{\varepsilon_1} - \frac{k_3}{\varepsilon_3}} \qquad 6.25$$

It is worth mentioning for infinite thickness ($a \to \infty$) (23) reduces to:

$$\frac{k_2}{k_2} = -\frac{\varepsilon_2}{\varepsilon_1} \qquad 6.26$$

This is the equation of two uncoupled surface plasmon polaritons at the respective interfaces.

We will from this point onwards consider the interesting special case where sub- and super substrate (II) and (III) are equal in terms of their dielectric response, i.e. $\varepsilon_2 = \varepsilon_3$ and $k_2 = k_3$. In this case the dispersion relation can be divided into a pair of equations namely:

$$\tanh k_1a = -\frac{k_2\varepsilon_1}{k_1\varepsilon_2}(a)$$
$$\tanh k_1a = -\frac{k_1\varepsilon_2}{k_2\varepsilon_1}(b) \qquad 6.27$$

It can be shown that Eq. (6.27a) describes modes of odd vector parity ($E_x(z)$ is odd, $H_y(z)$ and $E_z(z)$ are even functions); while Eq. (6.27b) describes modes of even vector parity ($E_x(z)$ is even function, $H_y(z)$ and $E_z(z)$ are odd).

The dispersion relations can now simulated with emphasis placed on the fact that as we annealed the sample, the thickness of the polymer layer reduces as evident by the x-ray results.

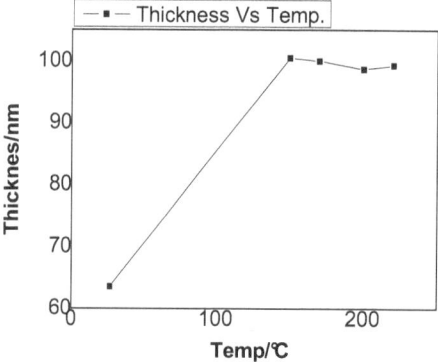

Figure 6.30: Thickness variation during the annealing process for colloids of diameter 60nm and 100nm

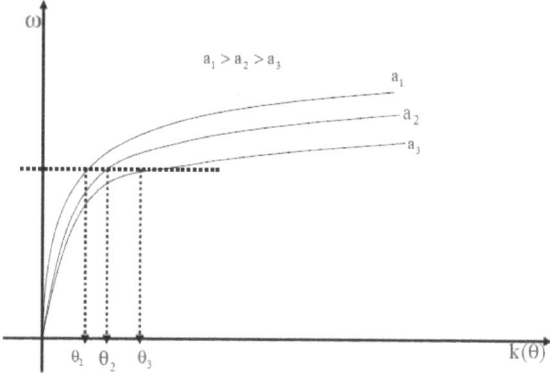

Figure 6.31: Dispersion relation of modes for an Au/polymer/Au multi layer. Notice the shift towards higher angles as a (polymer film thickness) reduces.

For simplicity here but without loose of generality the dielectric function of gold is approximated via a Drude's model with negligible damping ($\varepsilon(\omega)$ real) so that $\text{Im}(\beta) = 0$.
Odd modes have frequencies ω_+ higher than the respective frequencies for a single interface surface plasmon polaritons, and the even modes lower frequencies ω_-. For large wave vector β (which are only achievable if $\text{Im}[\varepsilon(\omega)] = 0$), the limiting frequencies are:

$$\omega_+ = \frac{\omega_p}{\sqrt{1+\varepsilon_2}}\sqrt{1+\frac{2\varepsilon_2 e^{-2\beta a}}{1+\varepsilon_2}}$$
$$\omega_- = \frac{\omega_p}{\sqrt{1+\varepsilon_2}}\sqrt{1-\frac{2\varepsilon_2 e^{-2\beta a}}{1+\varepsilon_2}}$$

6.28

Odd modes have the interesting property that upon decreasing metal film thickness, the confinement of the coupled surface plasmons polariton to the metal film decreases as the modes evolves into a plane wave supported by the homogeneous dielectric environment. For real, absorptive metal describe via a complex $\varepsilon(\omega)$, this implies a drastically increased surface plasmons propagation length. The even modes exhibit the opposite behaviour-their confinement to the metal increases with decreasing metal film thickness, resulting in a reduction in propagation length.

With respect to our multi layer, we set $\varepsilon_2 = \varepsilon_2(\omega)$ as the dielectric function of the metal and ε_1 as dielectric of the the insulating core. From energy point of view, the most interesting mode is the fundermental odd mode of the system which does not exibith a cut-off for vanishing core layer thickness.

6.2.1.1 Simulation of X-Ray Measurements using XOP2

During the annealing procedure, although the colloids cannot diffuse through the polymer layer due to their relatively larger size with respect to the polymer layer, their immediate contact surface with the polymer increases. This increase in surface contact is proportional to the annealing temperature. The figure below illustrates this situation.

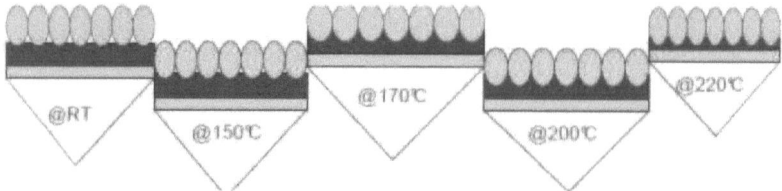

Figure 6.32: Increase of surface contact between polymer layer and 100nm gold colloids.

Considering at arbitrary temperature T > room temperature, we have colloids submerged almost halfway into the polymer layer and leaving the last section of the polymer layer untouched. In simulating such a layer system therefore we must take into account the fact that the surface of contact of the polymer layer with the colloids has a relatively higher electron density which decreases towards the base of the polymer layer. Since the colloids donnot totally displace the polymer layer and settle on the surface of the gold layer it further implies that there remains a thin polymer layer beneath with no contact with the gold colloids. Under this consideration, the polymer layer is subdivided into nine sub layers. At the immediate surface were the contact with colloids is greatest, the density an well as the gold content is highest. These decrease as we descend the layer and at the base of the layer we have just a thin layer of polymer. The diagram below illustrates this situation.

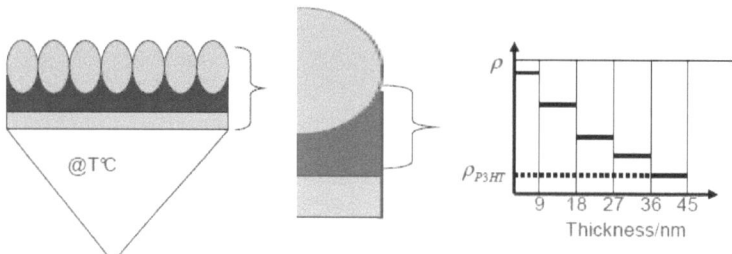

Figure 6.33: Simulation model showing variation of electron density within the polymer layer.

With the help of the simulation program XOP, we now simulate the layer structure by varying the aforementioned parameters until the simulation results fits our experimental findings.

Table 6.16: Simulation parameter at T=150°C

Thickness/nm	density	Electron density ratio(Au:P3HT)	Surface roughness
5	14	99.9:0.1	30
5	14	99.9:0.1	30
5	14	99.9:0.1	30
5	13	75:25	10
5	13	75:25	10
5	13	75:25	10
5	1.33	100%P3HT	20
5	1.33	100%P3HT	20
5	1.33	100%P3HT	20
60	19,85	100%Au	18

Table 6.17: Simulation parameter at T=170°C

Thickness/nm	density	Electron density ratio(Au:P3HT)	Surface roughness
5	14	99.9:0.1	50
5	14	99.9:0.1	50
5	14	99.9:0.1	50
5	13.8	99.9:0.1	10
5	13.8	99.9:0.1	10
5	13.8	99.9:0.1	10
5	1.33	100%P3HT	20
5	1.33	100%P3HT	20
4	1.33	100%P3HT	20
60	19,85	100%Au	17

Table 6.18: Simulation Parameter at T=200°C

Thickness/nm	density	Electron density ratio(Au:P3HT)	Surface roughness
5	14.4	99.9:0.1	50
5	14.4	99.9:0.1	50
5	13.9	99.9:0.1	50

5	13.7	99.9:0.1	10
5	13.7	99.9:0.1	10
3	13.7	99.9:0.1	10
5	1.33	100%P3HT	20
5	1.33	100%P3HT	20
60	19,85	100%Au	17

Table 6.19: Simulation parameter at T=220°C

Thickness/nm	density	Electron density ratio(Au:P3HT)	Surface roughness
5	14.4	99.9:0.1	30
5	14.4	99.9:0.1	30
5	13.6	99.9:0.1	30
5	13.6	75:25	10
3	13.6	75:25	10
5	1.33	100%P3HT	20
2	1.33	100%P3HT	20
60	19,85	100%Au	18

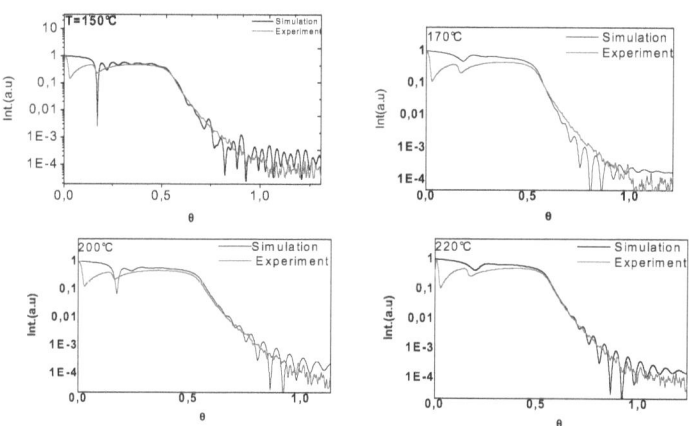

Figure 6.34: Simulation/Experimental results using the data from table 16-18.

The variation in the electron density within the polymer layer after each annealing step was then later plotted as a function of the layer thickness. This was calculated using the electron density ratio between polymer and gold. It is worth mentioning that an electron density ratio between gold/gP3HT of 99%:1% is obtained when we have 100 gold atoms to 1 polymer atom as well as when we have 2000 gold atoms to 1 gold atom. However the electron density variations of these two ratios are different. From the above tables, while we may see electron density ratios pretty similar, however the atomic constitutions are different. In order to avoid writing A/polymer complexes which do not exist, I resorted just in giving the ratios. However in calculating the electron density, the exact atomic constitution was taken into account.

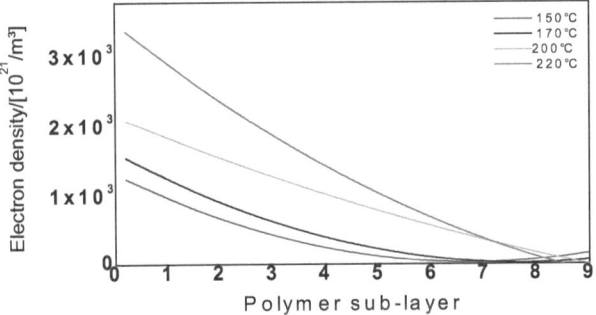

Figure 6.35: Variation in the electron density within the polymer layer as a function of temperature.

6.2.1.2 Simulation of SPR curves using x-ray model

Finally we would now use the calculated electron density from the x-ray results to simulate the SPR results. For each polymer sub layer, the corresponding real and imaginary component of the dielectric function is calculated using equation 6.15. With these values for each temperature, we now use the simulation program WINSPALL and simulate the layer system for each temperature.

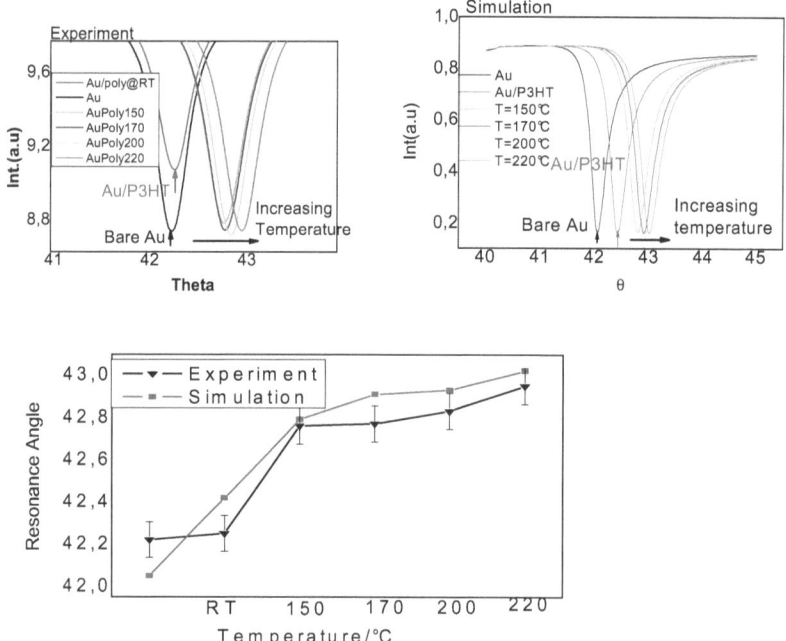

Figure 6.36: Experimental and simulation results. The third graphic compares the resonance angle at each given temperature.

With an experimental error of 2% we see again here that the experimental results are very consistent with the simulation results.

The device error was estimated by considering where errors are likely to occur in the course of the measurement. By considering figure 4.3, we see that the laser cannot be exactly attached to the zero degree strip of the goniometer due to the small difference in thickness between the attachment wire and two successive points of the goniometer. This difference is approximately 0,2mm

By running the scan from 0°-50°, we are therefore making an error within this range of 25mm. The experimental error therefore can be estimated using the expression:

$$\left(\frac{V_{exp.} - V_{abs.}}{V_{abs.}}\right) * 100\%.$$

6.2.1.3 Discussion

To understand the trend in the SPR response, we must again superimpose the evanescence wave field over the electron density variation within the polymer.

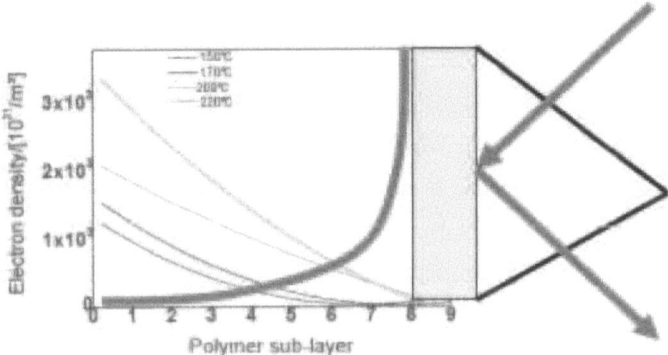

Figure 6.37: Evanescence wave field superimposed over the electron density variation within polymer sub-layers.

What one can observe is the fact that as the sample is annealed to higher temperatures, the colloids submerge into the thin polymer film. The submerging colloids create an electron density within the polymer film which decreases as one descends into the film. At T= 220°C, almost the entire 45 nm thin film has been submerged by the colloids hence the steep electron density variation within the polymer film at this temperature relative to the other temperatures. As one can see, a steeper electron density variation within the polymer layer has as a direct consequence a reduction of the net electron density that couples with the evanescence wave field.

Coupling at higher temperatures with a relatively lower electron density is seen on the SPR response as a higher dielectric constant (equation 3.1). This explains why varying the temperature leads to a movement of the coupling angle to the right.

7 Conclusion

Surface plasmon resonance sensing has been demonstrated in the past decade to be an exceedingly powerful and quantitative probe of the interactions of a variety of biopolymers with various ligands, biopolymers, and membranes, including protein:ligand, protein:protein, protein:DNA and protein: membrane binding. It provides a means not only for identifying these interactions and quantifying their equilibrium constants, kinetic constants and underlying energetics, but also for employing them in very sensitive, label-free biochemical assays [1-38].

In a typical SPR experiment, one interactant in the interactant pair (i.e., a ligand or biomolecule) is immobilized on an SPR-active gold-coated glass slide which forms one wall of a thin flow-cell, and the other interactant in an aqueous buffer solution is induced to flow across this surface, by injecting it through this flow-cell . When light (visible or near infrared) is shined through the glass slide and onto the gold surface at angles and wavelengths near the so-called "surface plasmon resonance" condition, the optical reflectivity of the gold changes very sensitively with the presence of biomolecules on the gold surface or in a thin coating on the gold. The high sensitivity of the optical response is due to the fact that it is a very efficient, collective excitation of conduction electrons near the gold surface. The extent of binding between the solution-phase interactant and the immobilized interactant is easily observed and quantified by monitoring this reflectivity change. An advantage of SPR is its high sensitivity without any fluorescent or other labeling of the interactants.

Physics however demands more than just the identification of a reaction. These demands with respect to this work were to investigate a process taking place within a thin film due to an external perturbation. Since the phenomena of SPR alone could not give us sufficient information to make a generalised conclusion on such a process, we combined our device with our home laboratory x-ray device. In this way the external perturbation could be investigated independently by these two spectroscopic methods.

These methods were necessary for the success of the above studies. They provided a means to convert local reflectivity changes measured in SPR microscopy to effective layer architectural changes and absolute surface coverage of adsorbed species determined by x-ray reflectivity. For a particular colloid diameter less than the thickness of the polymer film, the SPR coupling moved to lower angles. The vertically induced electron density variation within the film as well as its spatial re-localisation as a result of the thermal perturbation were found to be the cause of this non- linear peak movement. Further this induced electron density within the polymer layer was confirmed independently using x-ray reflectivity.

Although both spectroscopic methods were employed independently, the physical quantities which they detect i.e. electron density variation (x-ray reflectivity) and changes in the dielectric constant (SPR) are linearly dependent. As a consequence therefore knowledge of the electron density within the polymer layer at any given temperature can help us work backward the dielectric constant of the polymer layer at that given temperature and hence control the results we obtain from the SPR response.

By simulating the experimental findings obtained from x-ray reflectivity therefore the variation of the induced electron density within the entire polymer layer was calculated for every 5nm thickness of the polymer film. This then guarded us into calculating the dielectric constant of each sub layer at each of the temperatures.

An independent simulation was then carried out with these values and we found that the results were absolutely consistent with the response obtained from the SPR spectroscopy. The same analysis was done in the case where the diameter of the colloids does not enable their diffusion into the polymer layer.

Prior to these dynamic experiments, static measurements were carried as controlled experiments to investigate the functionality of our self made device. We also found out that all the results reproduced by the static analysis were consistent with values found in literature.

8 Bibliography

[1] Caruso, A. F. C. a. F., Biosensors: recent advances *Reports on Progress in Physics* **1997**, 60, (11), 1397-1445.

[2] Liebermann, T.; Knoll, W., Surface-plasmon field-enhanced fluorescence spectroscopy. *Colloids and Surfaces A: Physicochemical and Engineering Aspects* **2000**, 171, (1-3), 115-130.

[3] Neumann, T.; Johansson, M. L.; Kambhampati, D.; Knoll, W., Surface-plasmon fluorescence spectroscopy. *Advanced Functional Materials* **2002**, 12, (9), 575-586.

[4] Kwon, S. H.; Hong, B. J.; Park, H. Y.; Knoll, W.; Park, J. W., DNA-DNA interaction on dendron-functionalized sol-gel silica films followed with surface plasmon fluorescence spectroscopy. *Journal of Colloid and Interface Science* **2007**, 308, (2), 325-331.

[5] Buckle, M.; Williams, R. M.; Negroni, M.; Buc, H., Real time measurements of elongation by a reverse transcriptase using surface plasmon resonance. *Proceedings of the National Academy of Sciences* **1996**, 93, (2), 889-894.

[6] Ligler, F. S.; Rabbany, S. Y., *Synthetic microstructures in biological research* Plenum Press, New York: 1992; p 67-75.

[7] Turner;, A. P.; Karube;, I.; Wilson, G. S., *Biosensors: Fundamentals and applications* Oxford University Press, New York: 1987.

[8] Lowe, C. R., An introduction to the concepts and technology of biosensors. *Biosensors* **1985**, 1, (1), 3-16.

[9] Buerk, D. G., *Biosensors*. Technomic Publishing AG, Lancaster, USA: 1992.

[10] Hall, E. A. H., *Biosensors*. Springer, Heidelberg: 1990.

[11]. Raether, H., *Surface Plasmon on Smooth and Rough Surfaces and on Gratings* Springer, Berlin: 1988.

[12] Levy, C. D. P.; Cocolios, T. E.; Behr, J. A.; Jayamanna, K.; Minamisono, K.; Pearson, M. R., Feasibility study of in-beam polarization of fluorine. *Nuclear Instruments and Methods in Physics Research Section A: Accelerators, Spectrometers, Detectors and Associated Equipment* **2007**, 580, (3), 1571-1577.

[13]] http://www.esrf.eu/UsersAndScience/Experiments/TBS/SciSoft

[14] Mykytczuk, N. C. S.; Trevors, J. T.; Leduc, L. G.; Ferroni, G. D., Fluorescence polarization in studies of bacterial cytoplasmic membrane fluidity under environmental stress. *Progress in Biophysics and Molecular Biology* **2007**, 95, (1-3), 60-82.

[15] Su, C.-C.; Nikaido, H.; Yu, E. W., Ligand-transporter interaction in the AcrB

multidrug efflux pump determined by fluorescence polarization assay. *FEBS Letters* **2007**, 581, (25), 4972-4976.

[16] Tomin, V. I.; Oncul, S.; Smolarczyk, G.; Demchenko, A. P., Dynamic quenching as a simple test for the mechanism of excited-state reaction. *Chemical Physics* **2007**, 342, (1-3), 126-134.

[17] Physical Society of Japan, Journal (ISSN 0031-9015), vol. 54, Jan. **1985**, p. 278-281.

[18] Ramanavicius, A.; Kurilcik, N.; Jursenas, S.; Finkelsteinas, A.; Ramanaviciene, A., Conducting polymer based fluorescence quenching as a new approach to increase the selectivity of immunosensors. *Biosensors and Bioelectronics* **2007**, 23, (4), 499-505.

[19] Kumar, S.; Singh, P.; Kaur, S., A Cu^{2+} protein cavity mimicking fluorescent chemosensor for selective Cu^{2+} recognition: tuning of fluorescence quenching to enhancement through spatial placement of anthracene unit. *Tetrahedron* **2007**, 63, (47), 11724-11732.

[20] Uemura, T.; Furumoto, M.; Nakano, T.; Akai-Kasaya, M.; Saito, A.; Aono, M.; Kuwahara, Y., Local-plasmon-enhanced up-conversion fluorescence from copper phthalocyanine. *Chemical Physics Letters* **2007**, 448, (4-6), 232-236.

[21] Kiraz, A.; Doganay, S.; Kurt, A.; Demirel, A. L., Enhanced energy transfer in single glycerol/water microdroplets standing on a superhydrophobic surface. *Chemical Physics Letters* **2007**, 444, (1-3), 181-185.

[22] K. A Peterlinz and R. Geogianalis Optical Communication, 130 (4-6) 260-266 **1996**

[23] W Knoll, In R.E Hummel, P. Wissmann, (Eds) Handbuch of Optical Properties Band II, Optics of small particles, Interfaces and surfaces 373-399 CCR Press Boca Raton, New York, London, Tokyo, **1997**

[24] Als-Nielsen, Jens; McMorrow, Des: Elements of Modern X-Ray Physics. John Wiley and son (**2001**)

[25] G. J. Sprokel, R. Santo, J. D. Swalen, Mol. Cryst. Liq. Cryst 68 (**1981**) 16.

[26] G. J. Sprokel, Mol. Cryst. Liq. Cryst. 68 (1981) 39.

[27] U. Pietsch, T. A. Barberka, Th. Gueu and R. Strömmer, Il Nuovo Cimento Vol. 19D, N.2-4.

[28] Ma, Q.; Su, X.-G.; Wang, X.-Y.; Wan, Y.; Wang, C.-L.; Yang, B.; Jin, Q.-H., Fluorescence resonance energy transfer in doubly-quantum dot labeled IgG system. *Talanta* **2005**, 67, (5), 1029-1034.

[29] Graham, C. R.; Leslie, D.; Squirrell, D. J., Gene probe assays on a fibre-optic

evanescent wave biosensor. *Biosensors and Bioelectronics* **1992**, 7, (7), 487-493.

[30] Heido Nitta, Shigeru Shindo, Mitsuo Kitagawa, Surface Science 286, (**1993**) 346-354.

[31] Steinem, C.; Janshoff, A.; Ulrich, W.-P.; Sieber, M.; Galla, H.-J., Impedance analysis of supported lipid bilayer membranes: a scrutiny of different preparation techniques. *Biochimica et Biophysica Acta (BBA) - Biomembranes* **1996**, 1279, (2), 169-180.

[32] Sackmann, E., Supported Membranes: Scientific and Practical Applications. *Science* **1996**, 271, (5245), 43-48.

[33] Cornell, B. A.; BraachMaksvytis, V. L. B.; King, L. G.; Osman, P. D. J.; Raguse, B.; Wieczorek, L.; Pace, R. J., A biosensor that uses ion-channel switches. *Nature* **1997**, 387, (6633), 580-583.

[34]. Stelzle, M.; Weissmueller, G.; Sackmann, E., On the application of supported bilayers as receptive layers for biosensors with electrical detection. *J. Phys. Chem.* **1993**, 97, (12), 2974-2981.

[35] Hickel, W.; Knoll, W., Surface plasmon microscopy of lipid layers. *Thin Solid Films* **1990**, 187, (2), 349-356.

[36]. Corn, R. M.; Smith, E. A.; Wegner, G. J.; Goodrich, T. T.; Lee, H. J., SPR imaging measurements of DNA, peptide and protein microarrays. *Abstracts of Papers of the American Chemical Society* **2002**, 224, U151-U151.

[37] Lee, H. J.; Li, Y.; Wark, A. W.; Corn, R. M., Enzymatically Amplified Surface Plasmon Resonance Imaging Detection of DNA by Exonuclease III Digestion of DNA Microarrays. *Anal. Chem.* **2005**, 77, (16), 5096-5100.

[38] Fang, S.; Lee, H. J.; Wark, A. W.; Kim, H. M.; Corn, R. M., Determination of ribonuclease H surface enzyme kinetics by surface plasmon resonance imaging and surface plasmon fluorescence spectroscopy. *Analytical Chemistry* **2005**, 77, (20), 6528-6534.

[39] Hickel, W.; Kamp, D.; Knoll, W., Surface-plasmon microscopy. *Nature* **1989**, 339, (6221), 186-186.

[40] Porter, M. D.; Bright, T. B.; Allara, D. L.; Chidsey, C. E. D., Spontaneously organized molecular assemblies. 4. Structural characterization of n-alkyl thiol monolayers on gold by optical ellipsometry, infrared spectroscopy, and electrochemistry. *J. Am. Chem. Soc.* **1987**, 109, (12), 3559-3568.

[41] Whitesides, G. M.; Laibinis, P. E., Wet chemical approaches to the characterization of organic surfaces: self-assembled monolayers, wetting, and the physical-organic chemistry of the solid-liquid interface. *Langmuir* **1990**, 6, (1), 87-96.

[42] Bain, C. D.; Troughton, E. B.; Tao, Y. T.; Evall, J.; Whitesides, G. M.; Nuzzo, R. G.,

Formation of monolayer films by the spontaneous assembly of organic thiols from solution onto gold. *J. Am. Chem. Soc.* **1989,** 111, (1), 321-335.

[43] Nuzzo, R. G.; Dubois, L. H.; Allara, D. L., Fundamental studies of microscopic wetting on organic surfaces. 1. Formation and structural characterization of a self-consistent series of polyfunctional organic monolayers. *J. Am. Chem. Soc.* **1990,** 112, (2), 558-569.

[44] Lipshutz, R. J.; Fodor, S. P. A.; Gingeras, T. R.; Lockhart, D. J., High density synthetic oligonucleotide arrays. *Nature Genetics* **1999,** 21, 20-24.

120

[45] Lockhart, D. J.; Winzeler, E. A., Genomics, gene expression and DNA arrays. *Nature* **2000,** 405, (6788), 827-836.

[46] Nakatani, K.; Sando, S.; Saito, I., Scanning of guanine-guanine mismatches in DNA by synthetic ligands using surface plasmon resonance. *Nature Biotechnology* **2001,** 19, (1), 51-55.

[47] Muhlberger, R.; Robelek, R.; Eisenreich, W.; Ettenhuber, C.; Sinner, E. K.; Kessler, H.; Bacher, A.; Richter, G., RNA DNA discrimination by the antitermination protein NusB. *Journal of Molecular Biology* **2003,** 327, (5), 973-983.

[48] Parrat, L.G.: Surface Studies of Solid by Total Reflection of X-Rays. Phys. Rev. 95 **(1954)**

[49] Chechik, V.; Crooks, R.M.; Stirling, C. *Adv. Mater.* **2000,** *12,* 1161. Reactions and reactivity in self-assembled monolayers.

[50] a). Wang, J.; Pamidi, P.V.A.; Zanette, D.R. *J. Am. Chem. Soc.* **1998***, 120,* 5852. Selfassembled silica gel networks. b). Thompson, W.R., Cai, M.; Ho, M.; Pemberton, J.E. *Langmuir,* **1997,** *13,* 2291. Hydrolysis and condensation of self-assembled monolayers of (3-mercaptopropyl)trimethylsiloxane on Ag and Au surfaces.

[51] Journal of Molecular Structure Volume 218, March 1990, Pages 345-350

[53] Phys. Rev. 34, 1483–1490 (1929)

[54] www.biacore.com

[55] Analytica Chimica Acta Volume 307, Issues 2-3, 30 May 1995, Pages 253-268

[56] http://www.mpip-mainz.mpg.de/groups/knoll/

[57] http://www.esrf.eu/UsersAndScience/Experiments/TBS/SciSoft

[58] John David Jackson, Classical Electrodynamics, Third Edition.

[59] D. Axelrod, T. P. Burghardt and N. L. Thompson. Annual review in Biophysics and Bioengineering 13 247 – 268, **1984**.

[60] Notes on the classical Drude model by Kyle MCelroy

9 Acknowledgements

First and foremost, I would like to offer my sincere gratitude to my direct supervisor Prof. Ulrich Pietsch who has supported me throughout my thesis with his patience and knowledge.

I would also like to express my deep thanks to the co-supervisor Prof. Holger Shönherr for accepting to read through this work and to attest it.

For the construction of the surface plasmons resonance device and the subsequent adaption to our x-ray device I would like to extend my deep gratitudes to Tobias Panzner, G .Schmidt and his co-workers at the work shop of the physics department at the University of Siegen.

Special thanks also go to Herrn Gregor Schulte at the chemistry department of the University of Siegen for preparing the samples particularly the deposition of gold films on the surface of the prisms.

I would also like to extend my gartitudes to the co-workers at the University of Bochum for the Self Assembled Monolayers they prepared for us.

This entire work could not have been possible without financial assistance from the Siegener Graduitenkollege. In this respect my particular thanks go to the president, Prof. Dr. Michael Schmittel .

Last but not the least I would like to extend my entire gratification to all the members of the work group Festkörperphysik at the University of Siegen who directly or indirectly help me in the course of this thesis.

10 Affidavit
Eidesstattliche Erklärung:

Ich versichere an Eides Statt durch meine eigenhändige Unterschrift, dass ich die vorliegende Arbeit selbstständig und ohne fremde Hilfe angefertigt habe. Alle Stellen, die wörtlich oder dem Sinn nach auf Publikationen oder Vorträgen anderer Autoren beruhen, sind als solche kenntlich gemacht. Ich versichere außerdem, dass ich keine andere als die angegebene Literatur verwendet habe. Diese Versicherung bezieht sich auch auf alle in der Arbeit enthaltenen Zeichnungen, Skizzen, bildlichen Darstellungen und dergleichen. Die Arbeit wurde bisher keiner anderen Prüfungsbehörde vorgelegt und auch noch nicht veröffentlicht.

Siegen, 12.11.2010　　　　　　　　　　　　　　Felix Ntui Ayuk

Die VDM Verlagsservicegesellschaft sucht für wissenschaftliche Verlage abgeschlossene und herausragende

Dissertationen, Habilitationen, Diplomarbeiten, Master Theses, Magisterarbeiten usw.

für die kostenlose Publikation als Fachbuch.

Sie verfügen über eine Arbeit, die hohen inhaltlichen und formalen Ansprüchen genügt, und haben Interesse an einer honorarvergüteten Publikation?

Dann senden Sie bitte erste Informationen über sich und Ihre Arbeit per Email an *info@vdm-vsg.de*.

Sie erhalten kurzfristig unser Feedback!

VDM Verlagsservicegesellschaft mbH
Dudweiler Landstr. 99   Telefon +49 681 3720 174
D - 66123 Saarbrücken   Fax   +49 681 3720 1749
www.vdm-vsg.de

Die VDM Verlagsservicegesellschaft mbH vertritt

Die VDM Verlagsservicegesellschaft sucht für wissenschaftliche Verlage abgeschlossene und herausragende

Dissertationen, Habilitationen, Diplomarbeiten, Master Theses, Magisterarbeiten usw.

für die kostenlose Publikation als Fachbuch.

Sie verfügen über eine Arbeit, die hohen inhaltlichen und formalen Ansprüchen genügt, und haben Interesse an einer honorarvergüteten Publikation?

Dann senden Sie bitte erste Informationen über sich und Ihre Arbeit per Email an *info@vdm-vsg.de*.

Sie erhalten kurzfristig unser Feedback!

VDM Verlagsservicegesellschaft mbH
Dudweiler Landstr. 99
D - 66123 Saarbrücken

Telefon +49 681 3720 174
Fax +49 681 3720 1749

www.vdm-vsg.de

Die VDM Verlagsservicegesellschaft mbH vertritt

Printed by Books on Demand GmbH, Norderstedt / Germany